Melanie Schwab

A Decentralized Control Strategy for High Density Material Flow Systems with Automated Guided Vehicles

WISSENSCHAFTLICHE BERICHTE

Institut für Fördertechnik und Logistiksysteme
am Karlsruher Institut für Technologie (KIT)

BAND 85

A Decentralized Control Strategy for High Density Material Flow Systems with Automated Guided Vehicles

by
Melanie Schwab

KIT Scientific Publishing

Dissertation, Karlsruher Institut für Technologie (KIT)
Fakultät für Maschinenbau, 2015
Referenten: Prof. Dr.-Ing. K. Furmans, Prof. Dr. Kevin Gue

Impressum

KIT Scientific Publishing

Karlsruher Institut für Technologie (KIT)
KIT Scientific Publishing
Straße am Forum 2
D-76131 Karlsruhe

KIT Scientific Publishing is a registered trademark of Karlsruhe Institute of Technology. Reprint using the book cover is not allowed.

www.ksp.kit.edu

Print on Demand 2015

ISSN 0171-2772
ISBN 978-3-7315-0389-7
DOI: 10.5445/KSP/1000047227

A Decentralized Control Strategy for High Density Material Flow Systems with Automated Guided Vehicles

Zur Erlangung des akademischen Grades eines

Doktors der Ingenieurwissenschaften

der Fakultät für Maschinenbau
des Karlsruher Instituts für Technologie (KIT)
genehmigte

Dissertation

von

Dipl.-Wi.-Ing. Melanie Schwab

Tag der mündlichen Prüfung: 24.04.2015
Hauptreferent: Prof. Dr.-Ing. K. Furmans
Korreferent: Prof. Dr. Kevin Gue

Danksagung

Die vorliegende Arbeit entstand während meiner Tätigkeit als wissenschaftliche Mitarbeiterin am Institut für Fördertechnik und Logistiksysteme des Karlsruher Instituts für Technologie und wurde durch ein Auslandsstipendium des Karlsruhe House of Young Scientists (KHYS) unterstützt.

Ich bedanke mich herzlich bei meinem Doktorvater Prof. Dr.-Ing. Kai Furmans für die Betreuung meiner Arbeit und die konstruktiven Diskussionen. Außerdem bin ich sehr dankbar, dass er mich auf das Thema meiner Arbeit in Zusammenarbeit mit Prof. Dr. Kevin Gue aufmerksam gemacht hat und meinen Auslandsaufenthalt gefördert und unterstützt hat.

Ich bedanke mich bei Prof. Dr. Kevin Gue für die Übernahme des Koreferats und die Einladung zum Auslandsaufenthalt in Auburn 2012. Der Aufenthalt und die motivierenden Diskussionen haben entscheidend zum Gelingen der Arbeit beigetragen.

Ich danke Prof. Dr.-Ing. Thomas Böhlke für die Übernahme des Prüfungsvorsitzes meiner Promotionsprüfung.

In möchte mich bei allen Kollegen des IFL für die gemeinsame Zeit, gute Anregungen und moralische Unterstützung bedanken.

Meiner Familie und meinen Freunden möchte ich für die Unterstützung und die aufmunternden Worte auf dem Weg bedanken.

Ein besonderer Dank gilt meinem Mann Christian, der mich immer unterstützt, ertragen und wieder aufgebaut hat, wenn ich gehadert habe.

Sternenfels, April 2015 Melanie Schwab

Kurzfassung

Heutzutage macht die Wartezeit in Puffern oder Lagern den größten Teil der Durchlaufzeit von Ladeeinheiten in Supply Chains aus. Aufgrund der notwendigen Investitionen sowie der geringen Flexibilität sind Regallager als Kurzzeitpuffer nur bedingt geeignet. Blocklager haben dagegen den Nachteil des begrenzten Zugriffs auf einzelne Ladeeinheiten.

Daher werden neue Systeme entwickelt. In der Literatur existieren mehrere Systeme, die auf der Idee der "Puzzle-Based Storage Systems" basieren, und eine Bereitstellung nach dem Prinzip Ware-zum-Mann bieten. Die Ladeeinheiten werden dabei auf Fördertechnikelementen oder auf dem Boden gelagert und in einem Feld mit hoher Dichte und wenigen Leerstellen angeordnet. Im Rahmen dieser Arbeit wird die Bezeichnung "GridFlow Systems" für diese Systeme vorgeschlagen und eine Klassifizierung erarbeitet. Zusätzlich werden Vorteile der Implementierung mit fahrerlosen Transportfahrzeugen (FTF) für Großladungsträger vorgestellt.

Da für Systeme mit FTF keine universelle Steuerungsstrategie existiert, wird in dieser Arbeit die dezentrale Strategie "LivePath" erarbeitet. Hierbei trifft jedes Fahrzeug seine Entscheidung über die nächste Bewegungen unabhängig von der Entscheidung der anderen Fahrzeuge, sondern nur basierend auf dem aktuellen Systemzustand. Daher werden keine Routen vorausberechnet und reserviert, sondern die Route wird in Echtzeit bestimmt. Die Strategie löst Deadlocks und Livelocks bei deren auftreten auf und stellt dadurch die Funktionalität sicher.

Um die Lösungsgüte der Strategie bewerten zu können, wurde eine Methode zur Bestimmung der optimalen Bewegungen erarbeitet. Der Vergleich der optimalen Lösung mit der Lösung der Strategie zeigt eine gute Lösungsgüte der Strategie für ein einzelnes Fahrzeug. Für mehrere

Fahrzeuge konnte aufgrund der hohen Lösungszeiten für die exakte Lösung keine Aussage gemacht werden.

Zusätzlich wurde die Strategie in einer Simulationssoftware implementiert und es wurden Aussagen über das Systemverhalten in Bezug auf die Systemparameter sowie über Richtlinien zur Systemgestaltung abgeleitet. Die Analyse eines Puffer- und Crossdock-Szenarios zeigte, dass eine stärker spezialisierte Strategie für eine effiziente Crossdock-Abwicklung benötigt wird. Für Produktionsumgebungen lieferte die Strategie LivePath angemessene Durchsätze.

Abstract

In today's supply chains, loads spend most of their time waiting in buffers or storage systems. Due to investment and flexibility reasons, racking systems might not be the best solution for short-term buffer systems. Block storages on the other hand, hold the disadvantage of limited access to individual loads.

Therefore, new systems are being developed. In literature there are several systems based on the idea of Puzzle-Based Storage Systems providing a part to picker access to each load. In these systems the loads are stored on conveyors or on the ground. They are arranged in a grid with a high density and few empty locations. We developed a taxonomy to classify those systems proposing the name 'GridFlow systems', and we presented advantages of GridFlow systems with automated guided vehicles (AGVs) for large loads.

For GridFlow systems with AGVs no universal control strategy is available. Therefore, we provide the decentralized strategy 'LivePath', i.e. each AGV takes its own decision on the next move based on the current system state. There is no preplanning and reservation of paths, but the path is determined 'live'. The strategy resolves deadlocks and livelocks on occurrence and therefore guarantees functionality.

To assess the solution quality of LivePath, we present a method to calculate the optimal movements. The comparison of optimal and LivePath solutions showed a good solution quality for a single AGV. Multiple AGVs could not be analysed due to the solution times for the exact solution.

Furthermore, we implemented the strategy in simulation software and derived insights on the system behavior in dependence of the system parameters as well as rules of good system design. Additionally, we analysed a buffer and a cross dock case. Analysis showed that a specialised strategy for the cross docking task is needed, as throughput was limited. For production environments, LivePath provided a suitable throughput.

Contents

1 Introduction

Today's sourcing, production, logistics, and demand processes cannot be synchronized perfectly due to fluctuations in customer demand, transportation, lot sizes as well as variability in process times. Therefore, systems need to be decoupled to handle the fluctuations. This may happen by reserve capacity, waiting by customers, or waiting by material (buffer stock). As reserve capacity is very expensive, and customer waiting time is not acceptable for most companies, we today see this problem handled by buffer stock in most cases. Thus, loads in material flow systems spend most of their time waiting in buffers or storage systems (Bartholdi and Hackmann 2011, p.5; Arnold and Furmans 2009, p.173). On the other hand, space in production and logistics facilities becomes ever more limited and has to be used efficiently (Gue 2006). This usage is determined by two factors:

1. Usage of heights: In big cities we see sky scrapers if available space is rare. In storage systems, high-bay racks are used to store a big number of items on small surface (Arnold and Furmans 2009, p.194). Alternatively, there can be multiple building levels. This, however, holds the disadvantage of vertical transportation, e.g. lifts.
2. Density of the system: In a high density system at least one load may not be directly accessed from an aisle. This means that an interfering load has to be moved to provide access. Double-deep pallet racks or block storages are examples for high density storage systems (Gue 2006).

Racking systems, such as automated high bay racks, can achieve a very good space and room usage, as they are able to combine both factors as the height can be used efficiently (e.g. high bay racks) and racks with with multiple storage depths (high density) are available (Gue 2006). However, due to high investments, they may not economically be used for small short term buffer and storage systems with a high turn over of material in hours or days (Bartholdi and Hackmann 2011, p.200). Additionally, they are very unflexible in respect to layout changes or addition/removal of storage locations, as the installation is hard to move, and the control systems are very complex (Bartholdi and Hackmann 2011, p.200).

Block storage does not use racks and is therefore much more flexible in terms of changing the system. This storage type has a higher density than most racking systems. This may compensate the poor usage of heights to a certain extend. Block storage works well for smaller short-term buffer, sorting or storage systems with a small storage time of the items, e.g. storages, buffers in production, or cross-docking. But the retrieval of individual loads may require the fork lift driver to relocate a lot of other loads to get to the requested load, as the density in those systems is very high and thus no free access to each unit is available. This results in big effort for manual processes.

To take away this disadvantage of the limited access with manual processes in block storages, Gue and Kim (2007) propose a Puzzle-Based Storage System for high density storages. The system is able to provide a part to picker access to each load by transporting the requested load to a defined I/O-Point. It works with a grid of locations each equipped with a right angle transfer conveyor. For big loads, e.g. pallets, this will result in a very unflexible and expensive system, as the required conveyors are fixed to their location and very expensive.

The system idea may be alternatively implemented with automated guided vehicles (AGVs) moving underneath the loads and transporting them in the grid. The loads will not be stored in racks, but on the ground. This results in a very flexible high density system for large loads. The concept 'GridFlow with AGVs' can be used for different tasks in the material flow such as storing, buffering, and sorting. It can be implemented for any rectangular layout and the throughput can be adjusted to the requirements by changing the system parameters, e.g. number of AGVs or number of empty locations.

1.1 Aim and Scope of this Thesis

Different system designs and extensions of the original proposal of Puzzle-Based Storage Systems are presented in literature (see Chapter 2.1). Thus, a taxonomy to classify the different systems is developed (see Chapter 3). Furthermore advantages of GridFlow systems with AGVs for big loads are presented (see Section 3.2).

To implement the system control for GridFlow systems with AGVs, movement strategies for the AGVs are needed which are able to handle every grid size, number of AGVs, and number of empty locations (at least one). There are strategies for systems with conveyors (Gue and Kim 2007, Gue and Furmans 2011, Gue et al. 2012, Gue et al. 2013) as well as strategies which work for defined setups with AGVs (Alfieri et al. 2012, Furmans et al. 2011). However, there is no universal strategy for systems with AGVs in literature.

Therefore, we provide the decentralized control strategy 'LivePath' which can handle each rectangular system setup with AGVs (see Chapter 4). The strategy rules to detect and resolve deadlocks as well as livelocks are discussed and evaluated in detail (see Chapter 5).

To assess the solution quality, we present a method to determine the optimal movements (see Chapter 6). We use the method to assess LivePath concerning the retrieval times by comparing the optimal solution with the solution gained by the LivePath strategy (see Chapter 7).

Afterwards, insights on system behaviour and on useful system design are derived from the implementation of LivePath in simulation software (see Chapter 8).

At last two cases, a buffer and a cross dock, are analysed to obtain the achievable throughput and to analyse the influence of the vehicle parameters (see Chapter 9).

2 High Density Material Flow Systems

Gue (2006) defines a storage system to have a high density if interfering loads in the system sometimes have to be moved to access the desired load. According to this definition, a single-deep rack does not have a high density, as all loads are accessible at all times. A double-deep rack, on the other hand, has a high density, as the load in the front position has to be moved to access the load in the back.

The same holds true for flow racks where loads are stored in lanes. Within a lane the loads move to the front by gravity. A lane normally only holds loads of one product, as the access to loads in the back is very limited (Bartholdi and Hackmann 2011, p.54).

Block storage systems also have a high density as direct access is limited to the loads on the top. A fork lift has to move interfering loads to gain access (Martin 2011, p.358f). The interfering loads might be moved to other locations within the block temporarily. Often, there are not enough open locations to do so, and the loads have to temporarily be put out of the system and rearranged after the requested load was accessed. Therefore block storage is mainly used in systems with a high number of loads of the same product (Bartholdi and Hackmann 2011, p.54; Martin 2011, p.358f).

Another system with high density is proposed by Gue and Kim (2007). They introduce the concept of Puzzle-Based Storage Systems (PBSS) for very high density storage systems. The idea is derived from the 15-puzzle shown in Figure 2.1.

Figure 2.1: 15-puzzle (Gue and Kim 2007)

2.1 Puzzle-Based Storage Systems with Conveyors

The loads in the system presented by Gue and Kim (2007) are not stored in racks. In contrast to block storage, they are not stored on the ground either, but on conveyors which enable 4-way movements of the loads in the horizontal level. The conveyors and loads are arranged in a rectangular grid with at least one unoccupied conveyor. The locations correspond to cells in a grid. An empty cell in the grid is denoted as escort as the empty cell is necessary to move the requested load to the defined I/O point. The empty cell escorts the load on its way there. Figure 2.2 shows an implementation for the U.S. Navy. The loads in this implementation are stored on movable platforms which can be transported in all 4 directions to enable load moves. The conveyors enable immediate load movement whenever the adjacent location is empty.

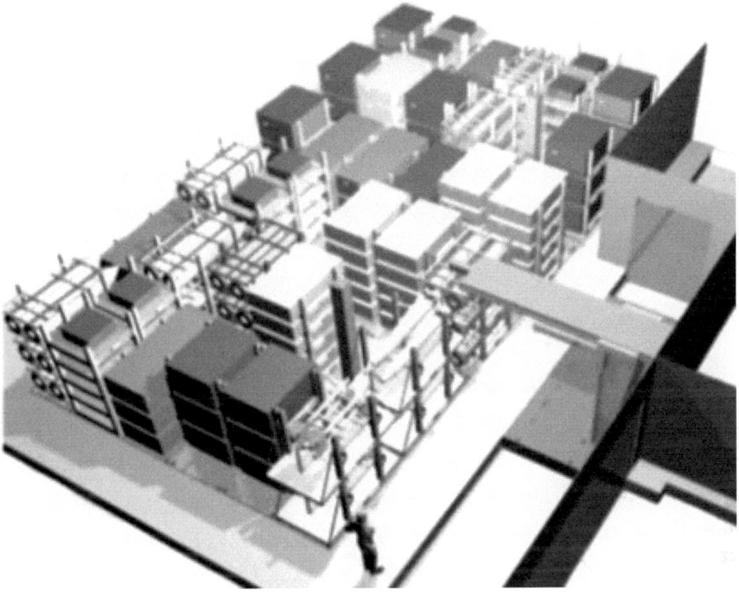

Figure 2.2: NavStore: a Puzzle-Based Storage System (Gue and Kim 2007)

2.1.1 Puzzle Strategy

Gue and Kim (2007) develop two algorithms based on puzzle moves for systems with one non variant I/O point.

In the algorithms three kinds of basic moves are used (see Fig. 2.3):

- 1-move: Moves the requested load to an adjacent escort with one conveyor movement
- 3-move: Moves the requested load to a position orthogonal to the adjacent escort with 3 conveyor movements
- 5-move: Moves the requested load to a position opposite to the adjacent escort with 5 conveyor movements

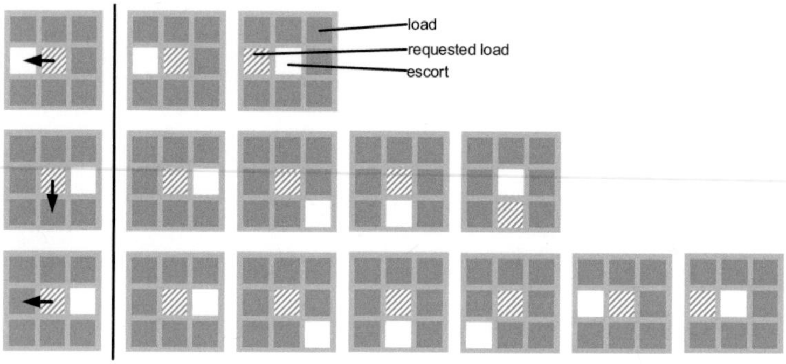

Figure 2.3: Types of load moves in single escort Puzzle-based storage system: 1-move (top series), 3-move (middle series), 5-move (bottom series) (following Gue and Kim (2007))

One algorithm works for a single escort system, the other works with multiple escorts. They present the algorithms and the corresponding analytical formulas for the retrieval times. In both algorithms the retrieval of a single load is considered, starting with a defined system state (escorts in defined positions at the I/O point). Replenishments and reestablishment of the initial system state are not included in the scope. Therefore, the retrieval time of one load can be found with the algorithms, but a long term throughput assessment with multiple simultaneous loads is not possible.

Gue and Kim (2007) find it to be beneficial to minimize the number of 5-moves. Instead 3-moves should be maximized by alternating the direction of load movement. Figure 2.4 shows an example illustrating the advantage of alternating change of direction, as four 3-moves are required instead of three 5-moves and one 3-move.

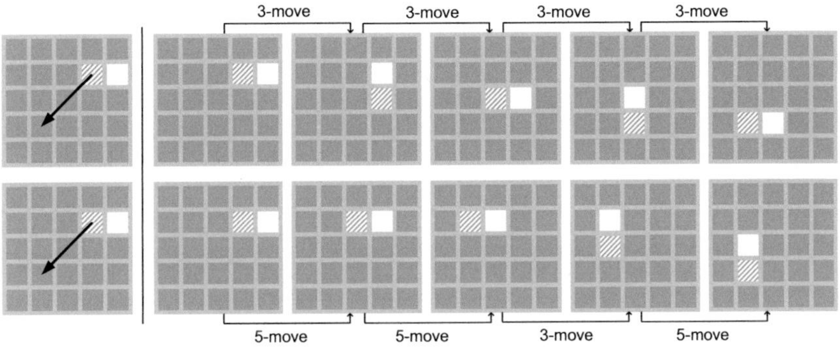

Figure 2.4: Top series: Alternating change of movement direction of the load, Bottom series: One change of direction of the load

Rohit et al. (2010) extend the algorithms by Gue and Kim (2007) for systems with one non variant I/O point. They give retrieval times for single escort systems with a randomly placed escort. Additionally, they give retrieval times for two escort systems with one escort being at the I/O point like in Gue and Kim (2007) and the other escort placed randomly in the system. Furthermore, they provide an integer program based on the Rush Hour problem. The problem considers one I/O point and multiple escorts without limitations to escort positions. The problem minimizes the number of load movements to retrieve one requested load from the system. Multiple requested loads or multiple I/O points are not considered.

Taylor and Gue (2009) extend this work by analysing the design and performance in multi-level Puzzle-Based Storage Systems and identifying potentials for implementation.

2.1.2 Virtual Aisle Strategy

Gue and Furmans (2011) propose the Virtual Aisle Strategy for Puzzle-Based Storage Systems with conveyors. In contrast to the puzzle strategy, there is not one I/O point, but all cells in the front row are output points in the system, and a requested load can leave the system at any of those cells. A requested load may only move towards the front row, and the loads blocking its way are cleared by moving right or left (see Fig. 2.5). The interfering loads form an aisle for the requested load to move through. The aisle will only be obtained while the requested load has not yet passed the concerned row.

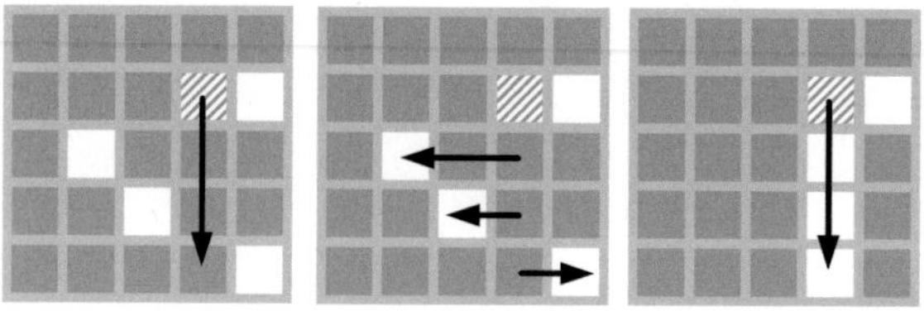

Figure 2.5: Exemplary retrieval movements of the Virtual Aisles strategy

At least one empty cell per row is needed for the strategy to work, as not requested loads may only move right or left. To obtain the number of empty cells in the grid a replenishment from the back aisle is initialized when a requested load leaves the grid at the front row.

In contrast to the puzzle algorithms (see Section 2.1.1), Gue and Furmans (2011) propose a decentralized control for the conveyors in the grid. The advantages of a decentralized control for this system are a higher robustness to breakdowns and a bigger flexibility for system changes (e.g. additional cells). Therefore, no centralized controller tells the conveyor which direction to convey to, but each conveyor takes this decision individually. Communication between conveyors to share information is implemented by messages that are passed on to the neighbor which passes the message further on if necessary. This approach is based on the algorithm by Mayer (2009) for a decentralized control of conveyors.

The so called Flexconveyors retrieve information about the system layout by messages that are distributed by all conveying units. Based on this knowledge, each conveying unit decides on the path the requested load on top of it, should take. After the decision, each conveying unit sends a request to the concerned conveyors on that path to do reservations. If multiple requested loads are in the system, the conveying units will negotiate paths. During the negotiation phase deadlocks and livelocks are prevented by detection of critical situations in the next move (Mayer 2009, Mayer and Furmans 2010). Gue and Furmans (2011) adapt this algorithm to implement the strategy Virtual Aisle.

The authors give sample results showing the throughput to increase with the number of empty cells per row as well as with the number of parallel requested loads.

Furmans et al. (2012) extend this work by providing insights on the failure behaviour of the system. They present a strategy to lessen the impact of a break down of single conveyor elements and to maintain the functionality of the system. This is enabled by the defined reactions of the conveyors to break downs of their neighbors (e.g. do no longer send loads to neighbors that do no longer answer the request, but reroute the load).

Gue et al. (2013) put the name 'GridStore' to this decentralized system with Virtual Aisle Strategy and present further insights on the system behavior as well as on design rules for GridStore systems. Furthermore, they present a proof of deadlock-freeness of the system rules.

2.1.3 GridPick

Gue and Uludag (2013) use the strategy Virtual Aisle to create a picking system called GridPick as an alternative to flow racks for small loads: "GridPick increases face density beyond one sku per slot by presenting only needed skus to the pick face, then withdrawing them to make room for upcoming picks. While the worker walks along the pick face picking the current order, items in the next order are making their way to the face in preparation for his next order. Ideally, workers never have to wait for an order not already at the face."

2.1.4 GridSequence

Gue et al. (2012) apply a puzzle-based system to the task of sequencing cartons. Storage systems may not be able to provide the required sequence for a palletizing robot if items are coming from different aisles, areas, or working stations. Therefore, a sequencing has to be established after picking the items. This task is fulfilled by a Puzzle-Based Grid of conveyors with the strategy Virtual Aisle located between picking and palletizing. Gue et al. (2012) present the system design and system rules, and give insights on system behaviour and efficient system design.

2.1.5 Sorting Strategy

Seibold et al. (2013) analyse different layouts for the FlexConveyor concept (see Section 2.1.2). They extend the algorithm presented by Mayer (2009) and analyse different layout options. The layout options include a rectangular grid. The authors derive insights on system behaviour from their analysis.

2.2 Puzzle-Based Storage Systems with Automated Guided Vehicles

Recently, a system implementation with AGVs has been proposed and investigated. The AGVs may move underneath the loads in the grid and transport the loads by lifting or dragging them, if they are equipped with wheels.

2.2.1 Puzzle Strategy

Alfieri et al. (2012) present an algorithm for multiple escort systems with one requested load, one AGV, and one fixed I/O point. The algorithm transfers the algorithm by Gue and Kim (2007) using the same load moves but enabling load movement by an AGV, instead of by conveyors. The authors present additional throughput data of case simulations with this algorithm.

Furmans et al. (2011) extend the algorithm by Gue and Kim (2007) and present analytical retrieval times for a system with one AGV, one escort, one requested load, and variable I/O point. The authors present an analysis of different positions for the I/O point and find that it should best be located in the middle of the longer side of the grid for the considered configuration assumptions.

Replenishments, the reestablishment of the starting state as well as multiple requested loads or multiple AGVs are not considered in both contributions. Therefore, the retrieval time of a single requested load can be found with the algorithms, but a long term throughput assessment is not possible.

3 GridFlow: Grid-Based Material Flow Systems

In literature, several different grid-based systems for different system tasks are presented (see Section 2.1 and 2.2). We propose the term 'GridFlow' as collective term for all grid-based material flow systems. We define GridFlow systems as automated grid-based material flow systems with very high density to convey, buffer, store, sort, or sequence loads.

The loads are arranged in a grid and can be moved in four directions. Sequential moves of loads allow the retrieval of requested loads that are not directly accessible. As no aisles are necessary to retrieve loading units, the grid-based layout offers a very high density. Empty cells in the system are called escorts (see Section 2.1). If there is at least one escort in the system, every load can be accessed. However, there might be other loads which have to be moved to retrieve a certain load. The fewer escorts in the system, the more movements have to be conducted to retrieve a certain load. This influences the retrieval times and, therefore, the throughput of the system. If many moves are necessary, retrieval times increase. Therefore, the operation point of the system can be chosen flexibly between the extremes:

- Very high density with one escort → longest retrieval time
- Space intense system with one loading unit → shortest retrieval time

3.1 Taxonomy of GridFlow Systems

The distinction between different GridFlow systems can involve the system task, the technical implementation, the system control as well as the movement strategy and system design (see Fig. 3.1).

GridFlow: Grid-Based Material Flow Systems				
System Task	**Loads**	**Technical Implementation**	**System Control**	**Movement Strategy and System Design**
• Buffer • Storage and Picking • Sorting • Sequencing	• Large loads • Small loads	• Steady material handling • Unsteady material handling	• Centralized • Decentralized	• Puzzle movements • Virtual Aisles • Flexconveyor algortihm • LivePath

Figure 3.1: Taxonomy of GridFlow systems

In literature, different system types have been presented, and different names have been coined. Table 3.1 classifies these systems according to the proposed taxonomy (see Tab. 3.1).

3.1.1 System Task

Up to now, authors mention storage, picking, sorting and sequencing as tasks for GridFlow system. Additionally, buffering for very short term storage is a possible task.

3.1.2 Technical Implementation

The system might be implemented with steady or unsteady material handling systems. So far, technical implementations with conveyors (e.g.

System	System Task	Loads	Technical Implementation	System Control	Movement Strategy and System Design
Gue and Kim 2007	Storage	not specified	Conveyors	Centralized	Puzzle Strategy
GridStore (Gue and Furmans 2011, Gue et al. 2013)	Storage	Small loads	Conveyors	Decentralized	Virtual Aisles
GridPick (Gue and Furmans 2011, Gue et al. 2013)	Picking	Small loads	Conveyors	not specified	Virtual Aisles
GridSequence (Gue and Furmans 2011)	Sequencing	Small loads (cartons)	Conveyors	Decentralized	Virtual Aisles
Alfieri et al. 2012	Storage and Picking	large loads (racks)	1 AGV	Centralized	Puzzle Strategy
Furmans et al. 2011	Storage	Large loads	1 AGV	Centralized	Puzzle Strategy
LivePath (this thesis)	Buffering, Storage and Picking, Sorting	Large loads	Multiple AGVs	Decentralized	LivePath

Table 3.1: Different GridFlow systems from literature

Flexconveyor proposed by Gue and Furmans (2011)) and implementations with AGVs have been proposed. While the technical details of the Flexconveyor are well described (Mayer 2009, Mayer and Furmans 2010), the work considering AGVs, up to now only discusses basic strategies for one AGV. Additionally, technical details have not yet been described (Nobbe (2015)). This thesis will focus on a strategy for a GridFlow system with AGVs. A detailed assessment of technical implementations can be found in Nobbe (2015).

3.1.3 System Control

System control might be implemented by a central controller which gathers all the information, takes decisions, and tells all the system elements what to do at all times. Alternatively, each system element may take its own decision decentralized. Decentralized control holds the advantage of flexibility and easy reconfiguration of the system (Gue and Furmans 2011). On the other hand, information handling is more difficult in a decentralized system, as information is stored and used for decisions in multiple system elements simultaneously. Information that needs to be shared with others has to be send via messages which may not arrive at all elements in time or not at all. Furthermore, decision taking is not coordinated in a decentralized system, and elements taking decisions simultaneously might lead to undesired system states. Decision taking and information sharing, therefore, have to be carefully designed in a decentralized system. In literature, there is a decentralized control for GridFlow systems with FlexConveyors presented, but no decentralized control for a system with AGVs.

3.1.4 Movement Strategy and System Design

In literature, rectangular grids have been discussed with different system designs according to the chosen movement strategy. Regarding the puzzle strategy there are multiple escorts lined up at the single I/O point. If AGVs are used with the puzzle strategy, all authors discuss implementations with only one AGV per grid.

The Virtual Aisles strategy requires at least one escort per row. Therefore, the system design depends to some extend on the chosen strategy and is hence matched with the movement strategy.

However, there are parameters of system design, such as grid size, that are not directly affected by the chosen movement strategy.

3.2 Gridflow Systems with AGVs

A GridFlow system may be implemented with conveyors (see Section 2.1) or with AGVs (see Section 2.2). For large loads, e.g. pallets or racks, the implementation with conveyors results in a very unflexible and expensive system, as the required conveyors are fixed to their location and very expensive. The implementation with AGVs offers a very flexible high density system for large loads.

Strategies for this implementation opportunity are not well discussed up to now. Only basic systems with one AGV have been described (Alfieri et al. 2012, Furmans et al. 2011). Therefore, this thesis will focus on a decentralized movement strategy for GridFlow systems with AGVs. We will assume AGVs that are moving underneath the loads and are able to transport the loads into all four directions in the horizontal level. Transverse movement is physically not possible. Figure 3.2 gives one technical implementation option complying with the given premises. For a detailed description of implementation options and their assessment see Nobbe (2015).

3.2.1 Advantages of GridFlow Systems with AGVs for Large Loads

Due to the automation of the system, it is less vulnerable to errors as human failure possibilities are minimized. The typical disadvantage of automated systems is the functionality of the system to depend on the functionality of each and every resource (Furmans et al. 2012). However, this holds not true for a GridFlow system with AGVs. If one vehicles fails the other vehicles will move around the vehicle until it is pulled out

Figure 3.2: Technical implementation option for a GridFlow system with AGVs (Nobbe 2015)

of the system by an employee. The failed vehicle will be fixed outside the system. During this time another vehicle will do its job.

The Gridflow system can be operated in a broad range of grid density and is scalable to bigger or smaller systems. The system hence is highly flexible to

- Throughput: The system can be adjusted to the desired throughput by adjusting the parameters, like the number of AGVs or the number of escorts (see Section 3.2.2), on an every day basis.
- Layout: Any layout that consists of interlinked grid locations is possible. However, this thesis will focus on rectangular grids as they are most applicable for industrial implementation.
- Paths: There is not one unique path through the grid a requested load has to take. On the contrary, the grid layout and the four possible movement directions offer the opportunity to transport loads on different paths.

A GridFlow system with AGVs for large loads requires little infrastructure compared to GridFlow systems with conveyors or other automated material flow systems. The area where AGVs move has to be separated to protect employees, the communication with the AGVs (e.g. by radio) has to be installed, and there needs to be a possibility to supply the AGVs with electricity (e.g. replacement batteries). Further infrastructure is not needed. Hence the system can be mounted and demounted quickly and does not need specific requirements.

The AGVs offer a good ratio of load capacity and vehicle weight, with the transported weight being bigger than the vehicle weight which helps to save energy. Additionally, the number of AGVs in the system is adjusted to the throughput and therefore no energy is wasted on unoccupied AGVs. Thus the system is energy-efficient.

The system costs can be classified to surface cost, investment costs, and operating costs. To summarize, these costs of a GridFlow system with AGVs are characterized by:

- low surface costs due to the high density of the system
- low investment costs compared to other automated systems as less fixed material handling equipment is required
- low labor costs compared to manual systems due to the automation
- low energy costs compared to other automated systems due to the energy-efficient characteristics of the system (see above)

The system costs depend strongly on the investment per AGV. The system can be universally applied for all tasks in material flow (conveying, buffering, sorting, storing). Therefore, a user can flexibly use his vehicles e.g. in different parts of his facility. This enables the production of big numbers of vehicles which makes production cost-efficient and decreases the price. Therefore, the system can be attractive to small and medium sized companies, too.

3.2.2 System Parameters

The system throughput is influenced by a number of system parameters and their interaction:

- size of the grid
- layout of the grid

- number of escorts
- number of vehicles
- number and positions of I/O-points
- movement strategy for retrieval orders and replenishments

This thesis aims to provide insights on the system behaviour of GridFlow systems with AGVs using the strategy LivePath (see Chapter 4) depending on the mentioned system parameters.

3.2.3 Modeling a GridFlow System with AGVs

As this thesis focuses on rectangular layouts, a GridFlow system with AGVs can be modeled by a grid with X columns and Y rows. The grid holds a total capacity of $N = X \cdot Y$ cells. Each cell can hold one load, but at least one cell in the grid has to be an escort for the system to work. Cells which do not hold a load are denoted as escorts (see Section 2.1.1). A grid is implemented on the floor of a facility. Therefore, there will be no vertical movement of vehicles. However in this thesis we will use the terms horizontal and vertical when denoting the two axes. As all figures in this thesis show top views of the system, it is intuitive to speak of vertical movements if talking about movement to the back or the front of the grid along the y-axe (see Fig. 3.3).

A position within the grid is identified by two coordinates x and y, where the x-coordinate stands for the 'horizontal' axe, while the y-coordinate is for the 'vertical' axe. Both axes start in the lower left corner with position (1,1) (see Fig. 3.3) (Gue and Kim 2007).

Grid elements might be vehicles, loads, and escorts. Loads and escorts might be assigned to an AGV. Figure 3.4 shows the symbols used in this thesis.

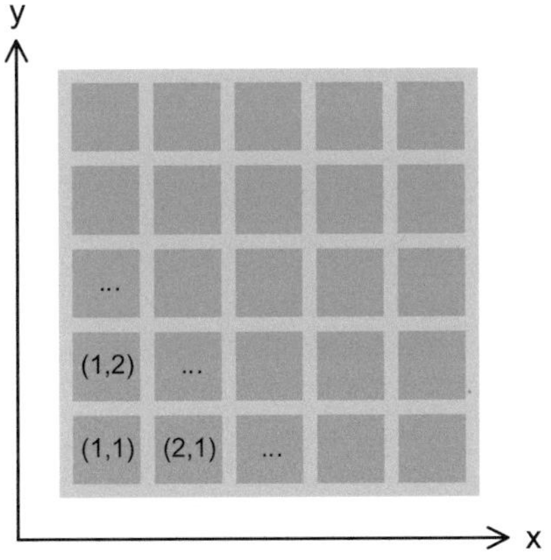

Figure 3.3: Basic layout of the system

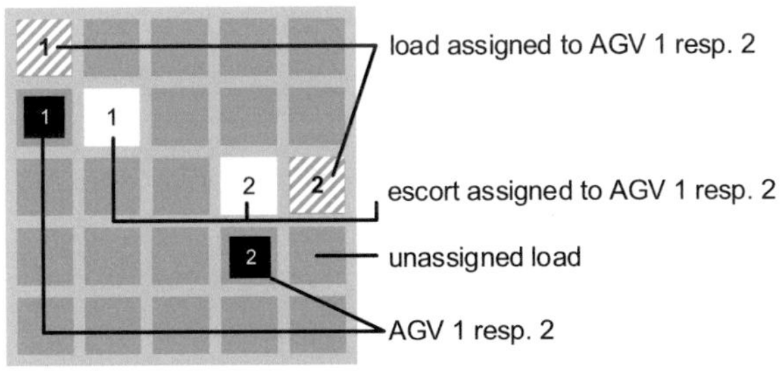

Figure 3.4: Symbols for grid elements (here: 2 AGVs)

4 Rule Set of the Strategy LivePath

As there is no universal strategy for systems with AGVs in literature, we developed the strategy LivePath. LivePath is a control strategy which can handle any number of escorts, AGVs, and parallel requests in a grid without limiting AGV movements to certain rows, columns or areas in the grid. Any number of I/O points can be handled, and the positions of the I/O points are not restricted to certain positions. All decisions are made from the individual point of view of each AGV without cooperation of AGVs. All AGVs determine their next move depending on the current system state which is known to all AGVs.

Each AGV is assigned to one load (requested or replenishment load) and considers the situation in the grid to find the next feasible move. In doing so, the planning and decision horizon of the AGVs differ from each other. The AGV might consider multiple future moves, thus the planning horizon might be several moves. But only the next AGV move is determined, thus the decision horizon is only the next AGV move. Therefore, there is no preplanning and reservation of paths, but the path is determined 'live', and only the decision on the next cell to move to is fixed.

An AGV might decide on the next move in two situations:

- The AGV is currently not moving (AGV has finished the previous move or has just entered the system)
- The AGV is currently moving and has to start the deceleration now, to stop in time before passing to the next cell. The decision is done in this situation to enable AGVs to avoid unnecessary deceleration and acceleration by moving on. The decision is taken as late as possible to decrease the risk of early fixations, which may lead to

unnecessary waiting or moving times due to interference by other AGVs in the meantime.

Each AGV takes its individual decisions. However, to do so the AGVs have to share certain information about the system state:

- Positions of all loads, AGVs, and escorts in the system
- AGV ranks (see Section 4.1)
- Assignments of escorts to AGVs
- Blocked cells due to fixated AGV moves (start and end cells)
- List of AGVs currently handling a deadlock or livelock situation (see Section 4.6)

This shared knowledge might be implemented by a central database where all AGV upload updates and get their information, or by individual databases in each AGV which are updated by messages that AGVs send to all others in case of changes. Therefore, the strategy LivePath potentially can be implemented to be completely decentralized. However, the implementation in simulation software in this thesis is not completely decentralized and described in Section 4.6.3.

4.1 General Assumptions

In the strategy LivePath several general assumptions are made because they are required for the system functionality or seem to offer benefits regarding the expected number of AGV moves.

AGV ranks

AGVs are assigned to ranks with 1 being the highest rank. No two AGVs may have the same rank. This assumption is needed for the deadlock and livelock handling in this strategy (see Section 4.6). The ranks might for example be assigned due to priorities of the requested loads, but AGVs with replenishment loads need to have a lower rank than those AGVs with a requested load. In this thesis, the initial ranks are chosen randomly. Whenever one AGV leaves the grid with its assigned requested load, all other ranks are updated to eliminate the rank gap. A new AGV that enters the grid will always have with the lowest rank of all AGVs.

Clear bigger distance first

Whenever an AGV defines the moving direction of its assigned load, it clears the bigger (horizontal or vertical) distance first. This transports the load faster into the neighborhood of the I/O point, thus reducing the probability of an unnecessary later move due to changes in the meantime (e.g. by other AGVs).

Alternate direction for loaded moves

The AGV will try to alternate horizontal and vertical moves if a vertical and horizontal distance has to be cleared by a load (see Section 2.1.1). This assumption supports the clearance of the bigger distance first, as it will automatically lead to the desired alternation of directions.

Do not alternate direction for unloaded moves

If determining the moving direction, of an unloaded AGV, the AGV will perform all moves in one direction first and then change the direction as this will reduce de-/acceleration and steering.

Prefer moves towards grid mid

If the AGV has to choose between left or right moves in horizontal resp. between up or down moves in vertical direction, it prefers moves towards the mid of the grid as moving options are smaller at the grid edge. This will only be followed if no decision from above apply.

Inconclusive situations

In other decision situations we cannot easily identify moves that promise clear benefits at all times. In those situations a random choice could be implemented. To reduce stochastic effects and thus, to facilitate the prediction of system behaviour, the assessment of deadlock behaviour, and the derivation of analytical estimations of system throughput, we fixate following additional assumptions though they do not provide an immediate benefit:

- Prefer horizontal moves to vertical moves
- Prefer moving to the right to moving to the left
- Prefer moving down to moving up

4.2 Decision Process of AGVs

The decision process of an AGV can be separated into five steps. It will fixate the next AGV move in this decision. Therefore, no step of the decision process includes actual AGV movement, but the move will be done after the decision process is finished.

1. Each AGV first determines the next position (next target cell) to move the assigned load to (see Section 4.3).
2. Next the AGV decides which escort to use to enable load movements (see Section 4.4).
3. The selected escort has to be moved to the next target cell of the assigned load to swap the load with the escort. This will move the assigned load to the next target cell. Therefore, the AGV determines the next move in the series of moves to enable this (see Section 4.5).
4. As all AGVs take their decisions individually this might result in blockings, deadlocks, or livelocks. Therefore, rules are implemented to force AGVs with a lower priority to wait or retreat if necessary (see Section 4.6).

The decision process of a single AGV is also given in decision trees in Appendix A.

4.3 Step 1: Find next Target Cell to move the Assigned Load to

In this step, the next target cell t_l for the assigned load l is determined. The cases of replenishment loads and requests have to be differentiated here.

4.3.1 Replenishment Load

If the assigned load is a replenishment load entering the grid, it will always be moved to the neighboring cell in the grid (see Fig. 4.1). Therefore, by selecting the waiting position of the replenishment load at the grid edge the next target cell t_l is defined. The rules to define the waiting positions in this thesis are described in Section 4.6.3.

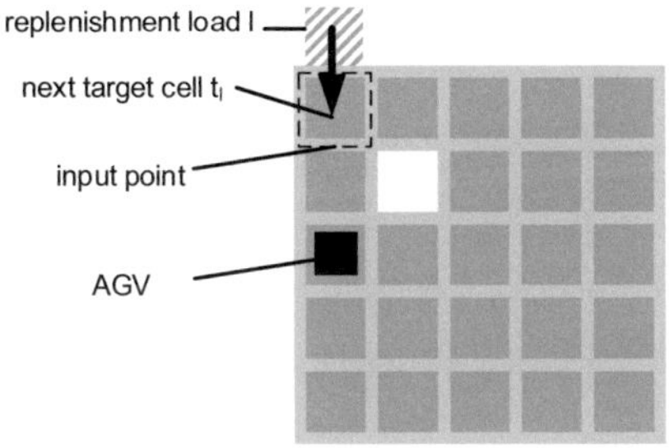

Figure 4.1: Next target cell for a replenishment load entering the grid

4.3.2 Requested load

If the assigned load is a requested load, all possible paths would have to be evaluated to find the optimal one. However, there are a lot of possible paths, and preplanning works poorly as other AGV will constantly change the situation for the considered AGV. Thus, we define a rule based on the current situation to determine the next target cell to move the load to. In this rule we consider the shortest distance of the requested load to one of the output points (see Fig. 4.2).

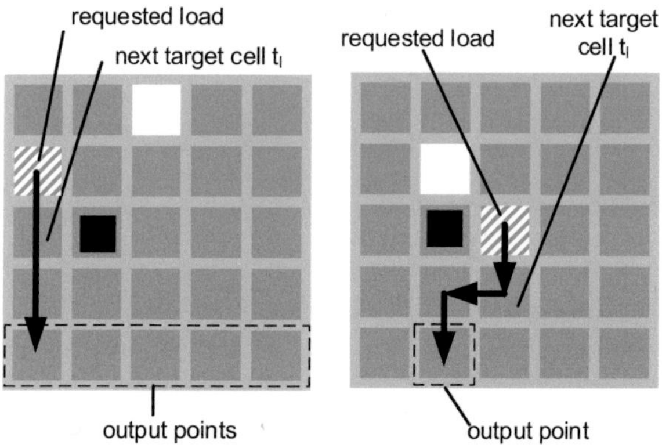

Figure 4.2: Next target cell for shortest distance of the requested load to the output point (left: example with multiple output points, right: example with single output point)

Shortest distance of requested load to output point

To determine the target cell $t_{l,s}$ for load l according to the shortest distance, the closest defined output cell for this load has to be found. To measure the distance, the rectangular metric is used to calculate the distance of load l to its output point op_l $d_{OP}(l, op_l)$. $d_{x,OP}(l, op_l)$ and $d_{y,OP}(l, op_l)$ denote the horizontal resp. vertical distance. To calculate the horizontal distance the absolute difference of the columns x_l and x_{op_l} of load l and output point op_l is used. The vertical distance is obtained accordingly.

$$d_{OP}(l, op_l) = d_{x,OP}(l, op_l) + d_{y,OP}(l, op_l) = |x_l - x_{op_l}| + |y_l - y_{op_l}| \quad (4.1)$$

In the single output point example of Fig. 4.2 the horizontal distance $d_{x,OP}(l, op_l) = 1$ and the vertical distance $d_{y,OP}(l, op_l) = 2$.

If there are multiple output points the closest one has to be chosen. The distance $d_{OP}(l)$ of load l to the closest output point is the minimal distance to all output points.

$$d_{OP}(l) = \min_{op_l} \left(d_{OP}(l, op_l) \right) \tag{4.2}$$

If there are multiple output points with an identical distance, the choice is made according to the general assumptions (see Appendix A).

To get to the selected output point op_l^* the box might be moved in horizontal or vertical direction towards the output point. Thus there are up to four possibilities for the next target cell. Depending on the distances $d_{x,OP}(l, op_l^*)$ and $d_{y,OP}(l, op_l^*)$ the following rules are chosen for the strategy LivePath (see also Appendix A):

- clear bigger distance ($d_{x,OP}(l, op_l^*)$ or $d_{y,OP}(l, op_l^*)$) first (see Section 4.1)
- else: If distances $d_{x,OP}(l, op_l^*)$ and $d_{y,OP}(l, op_l^*)$ are equal, move orthogonal to last move (Gue and Kim 2007 and Section 4.1).
- else: If there was no move before, but one of the alternatives for the next target cell is an escort, prefer this alternative.
- else: If no or all alternative cells are escorts, prefer to move towards the grid mid (see Section 4.1).
- else: If all alternative cells have the same distance to grid mid, prefer horizontal moves to vertical ones and moves to the right to left and down to up (see Section 4.1).

Figure 4.2 shows the paths that the loads will move on if the situation does not change. However only the next target cell is fixated, not the complete path. If the situation changes until the next move is finished (e.g. another AGV moves a load on the path), the path will be altered, i.e. the next target cell will be defined anew as stated above (see Chapter 4).

If the determined target cell with shortest distance $t_{l,s}$ is occupied, it might be beneficial to use escorts nearby. This could increase the length of the path of the requested load, but it might decrease the number of AGV moves and the number of all load moves for this retrieval. Therefore, after the target cell with shortest distance $t_{l,s}$ is found, options for fast partial paths are evaluated.

Fast Partial Paths for Requested Load

A fast partial path might be useful if $t_{l,s}$ has to be cleared first. Figure 4.3 gives an example for a useful fast partial path. The considered fast partial path options always contain one cell orthogonal and one cell parallel to the path of shortest distance. A fast partial path is considered to be useful if all of the following conditions are true:

- The next target cell $t_{l,s}$ for the shortest distance is occupied and has to be cleared first.
- Both cells of the fast partial path are escorts (escort can be assigned or unassigned to an AGV).
- After the load is moved to the second cell of the fast partial path the minimum distance $d_{OP}(l)$ to one of the output points is equal or decreased in comparison to the minimum distance calculated from $t_{l,s}$. The closest output cell op_l^* might change due to the fast partial path, but the minimum distance may not increase by using the fast partial path, compared to moving to $t_{l,s}$.

There might be up to four options for fast partial paths (see Fig. 4.4). All options are analyzed beginning with and preferring the ones towards the mid of the grid. If multiple options point towards the mid of the grid, the general assumptions apply (see Section 4.1 and Appendix A). If there is a useful fast partial path found, the corresponding target cell $t_{l,f}$ is determined according to the above rules and the next target cell $t_l = t_{l,f}$. If there is no useful fast partial path the next target cell $t_l = t_{l,s}$.

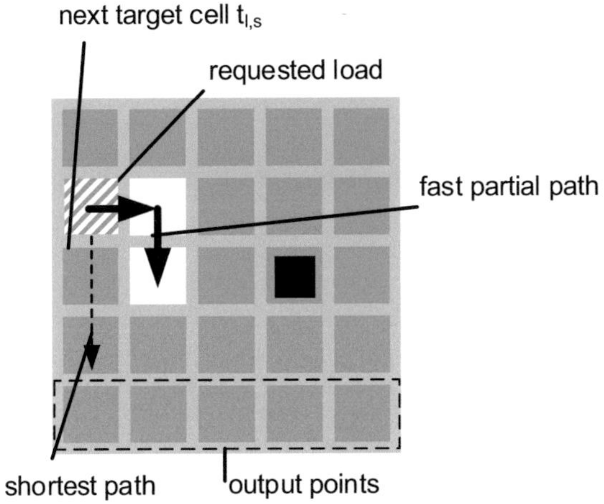

Figure 4.3: Example for a fast partial path for the requested load

4.4 Step 2: Find Escort to Use

To move the assigned load (requested or replenishment load) to the next target cell t_l, the AGV needs to move an escort into t_l in order to move the assigned load to t_l. To enable this, the AGV will choose one escort based on the current situation which it will then transport to t_l. This decision is done after each move, therefore the AGV will switch to another escort if a more promising escort is available.

As one escort may not be used by multiple AGVs at a time, each AGV can assign an escort to itself blocking other AGVs from using this empty cell. This information is part of the information shared by the AGVs (see Section 4.1).

However, an AGV may assign an escort of an AGV with lower rank to itself if the affected AGV is not currently moving at this time. The affected AGV then will reselect its escort before starting the next move. This will enable AGVs with a high rank to use the best escort and finish

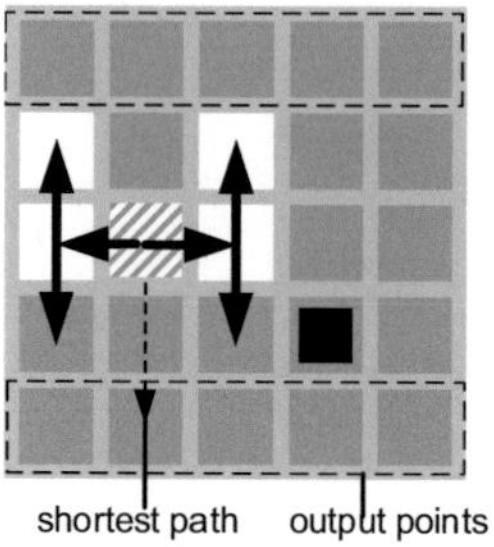

Figure 4.4: Possible fast partial paths options for a requested load

their retrievals fast. Escorts involved in current AGV moves to/from this cell cannot be assigned by an AGV determining its escort.

All assignable escorts for an AGV are considered and the number of AGV moves to transport the escort to the target cell t_l is estimated. For the estimation two phases have to be considered. First, the AGV has to move unloaded to the considered escort. Secondly, the AGV has to transport the escort to t_l. The escort with the smallest estimated number of AGV moves of both phases is selected.

Estimation Phase 1: Move the AGV to the Escort

First the AGV v has to move from its current position to the escort e (see Fig. 4.5). As the AGV is unloaded, it needs one move to go from one cell to the other. On the AGVs path interferences with other AGVs might be possible. Thus, the number of moves by the AGV $n_{ve}(e, v)$ cannot be determined exactly, but can be estimated by the lower bound:

$$n_{ve}(e, v) = |x_v - x_e| + |y_v - y_e| \tag{4.3}$$

Estimation Phase 2: Transport the Escort to the Target Cell

Next the AGV v has to move the escort e to the target cell t_l (see Fig. 4.6). It takes three AGV moves to transport the escort to an adjacent

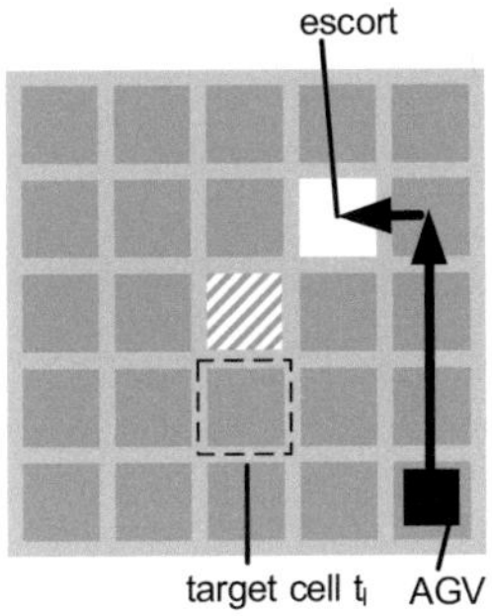

Figure 4.5: Estimation phase 1: move the AGV to the escort (unloaded moves)

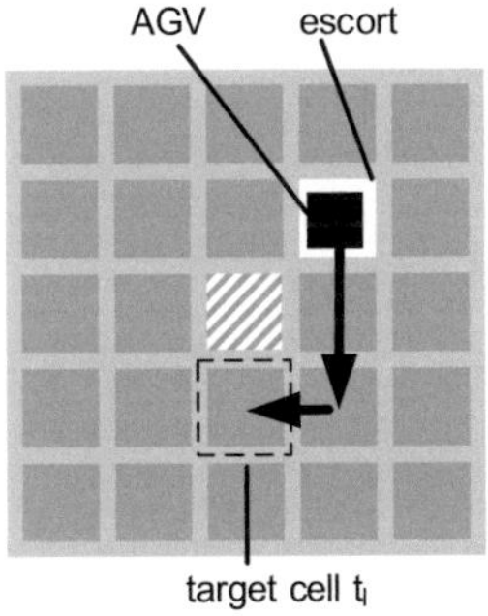

Figure 4.6: Estimation phase 2: move the escort to the target cell

occupied cell, with the AGV starting in the escort to be moved and in the end being in the moved escort (see Fig. 4.7).

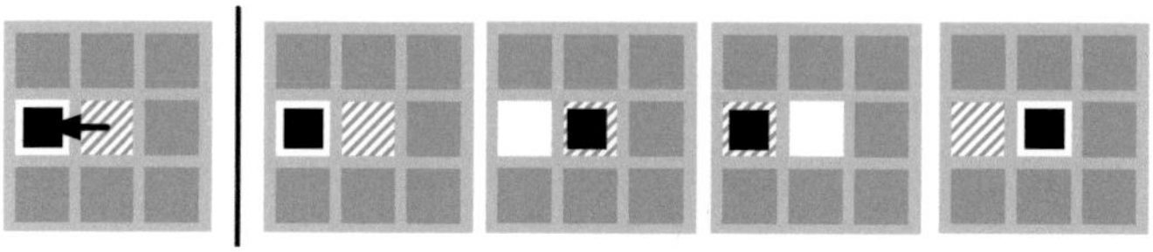

Figure 4.7: 3 moves to transport the escort by one cell

In this estimation we use a worst case scenario concerning the loads on the path, because possible escorts in the path might be used by other AGVs in the time it takes to get there. Therefore, all cells on this path are assumed to be occupied.

Furthermore, the AGV might not need to move into the escort when it reaches the escort or once the escort reaches the target cell t_l, but this is still assumed (worst case).

Additionally, the AGV might have to take additional moves due to other AGVs. As this is difficult to predict and a worst case assumption would result in a very big number of moves, this is not considered.

Thus, no useful lower or upper bound for the number of AGV moves $n_{etc,dir}$ to transport the escort to the target cell can be determined, but it can be estimated by:

$$n_{etc,dir}(e, v, t_l) = 3 \cdot (|x_e - x_{t_l}| + |y_e - y_{t_l}|) \tag{4.4}$$

A special case arises if the escort e, the assigned load l and the target cell t_l are in one line. Then a detour path is needed (see Fig. 4.8) in order not to involuntarily move the assigned load. This detour requires the escort to be moved for two additional cells compared to the hypothetical direct path. Thus, the detour requires six additional AGV moves (three AGV moves per cell) in the worst case scenario.

Again all cells on this detour path are assumed to be occupied (worst case), but due to possible interferences of other AGVs, there can no useful

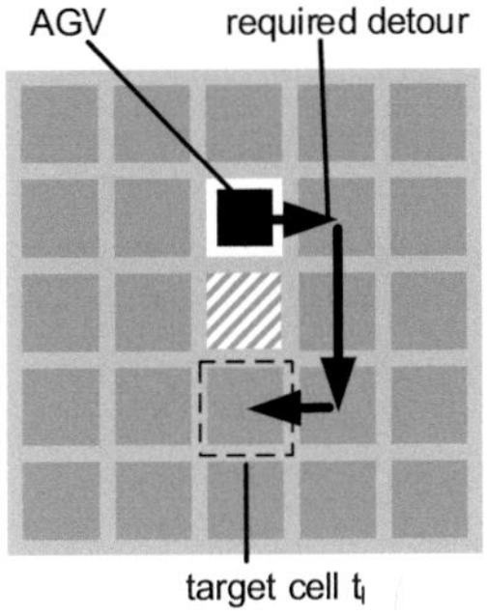

Figure 4.8: Estimation phase 2: Special case of escort, assigned load and target cell in one line

upper or lower bound determined. However, the number of additional AGV moves $n_{etc,det}$ for the detour can be estimated by:

$$n_{etc,det}(e, v, t_l) = \begin{cases} 6, & e,\ l \text{ and } t_l \text{ are in a line} \\ 0, & \text{else} \end{cases} \tag{4.5}$$

Combination of Phase 1 and 2

The total number of AGV moves $n_{etc,total}$ for phase 1 and 2 can be estimated by:

$$n_{etc,total}(e, v, t_l) = n_{ve}(e, v) + n_{etc,dir}(e, v, t_l) + n_{etc,det}(e, v, t_l) \tag{4.6}$$

This estimation is calculated for all assignable escorts and the escort with the minimal number of estimated AGV moves is chosen to be the assigned escort e_l. If there are multiple escorts with minimal number of moves the general assumptions apply (see Section 4.1 and Appendix A).

4.5 Step 3: Find Next AGV Move in Order to transport the Escort to the Target Cell

Step 1 determined the next target cell t_l to move the load to. Step 2 determined the escort e_l to use. In Step 3 the next AGV move (loaded or unloaded) is determined based on the current system state. Three cases have to be distinguished:

- requested load is in one of its output point
- escort e_l is in the target cell t_l
- escort e_l is not in the target cell t_l

4.5.1 Case 1: Requested Load is in one of its Output Points

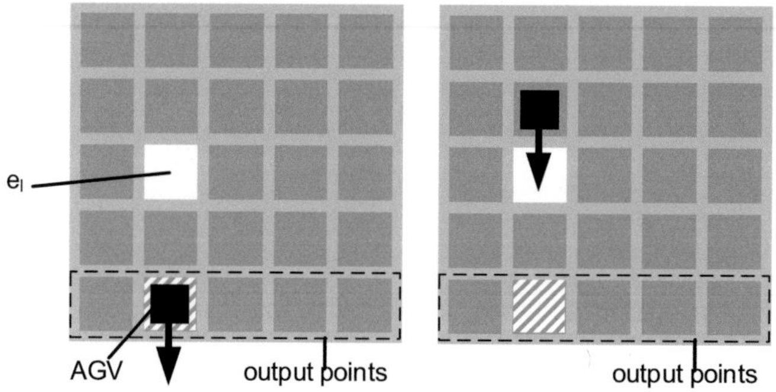

Figure 4.9: Case 1: requested load is in one of its output points (left: AGV under requested load, right: AGV not under requested load)

If the requested load is in one of its output points, it can be taken from the system. If the AGV is under the load, it moves the load from the system (see 4.9, left). If the AGV is not under the load, it moves towards the load position now (see Fig. 4.9, right) and will take the load from the grid when it reaches the load.

4.5.2 Case 2: Escort e_l is in the Target Cell t_l

This case can only apply if the assigned load is not yet in one of its input resp. output points. Requested loads and replenishments loads have to be distinguished.

Requested Load

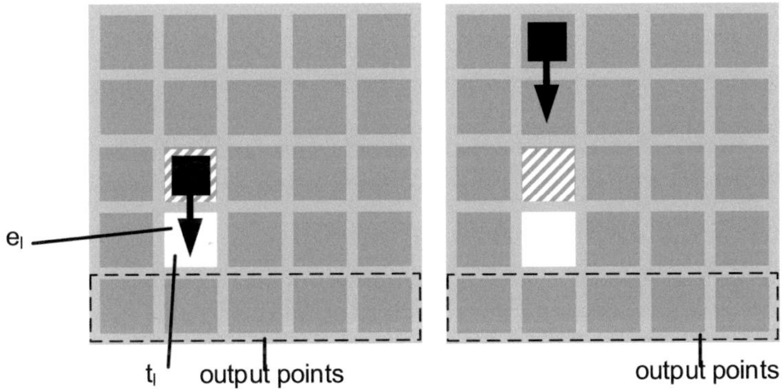

Figure 4.10: Case 2: escort e_l is in the target cell t_l (left: AGV under requested load, right: AGV not under requested load)

If the escort e_l is in the target cell t_l the load l can be moved to the target cell t_l. If the AGV is under the load l, it moves the load to the escort e (see Fig. 4.10, left). If the AGV is not under the load, it moves towards the load position (see Fig. 4.10, right) and will move the requested load when it reaches the load.

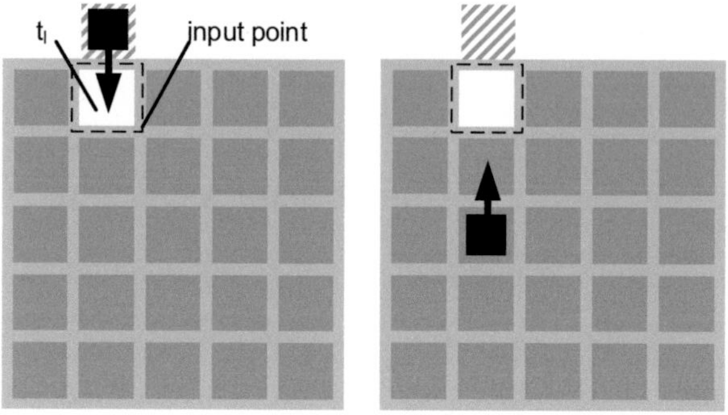

Figure 4.11: Case 2: escort e_l is in the target cell t_l (left: AGV under replenishment load, right: AGV not under replenishment load)

Replenishment Load

A replenishment load will only be moved once, to enter the grid. After this it will change its status and will no longer be assigned to an AGV. If the escort is in the target cell t_l and the AGV is under the replenishment load, the AGV moves the load into the grid (see Fig. 4.11, left). If the AGV is not under the load, it moves towards the load position now (see Fig. 4.11, right) and will move the replenishment load into the grid when it reaches the load.

4.5.3 Case 3: Escort e is not in the Target Cell t_l

This case can only apply if the assigned load is not yet in one of its input resp. output points (see Fig. 4.12). Requested loads and replenishment loads do not have to be distinguished in this case.

If the escort e_l is not yet in the target cell, it has to be transported there. Therefore, the next target cell t_e for the escort e_l has to be determined based on the target cell for the load t_l as well as the position of the escort.

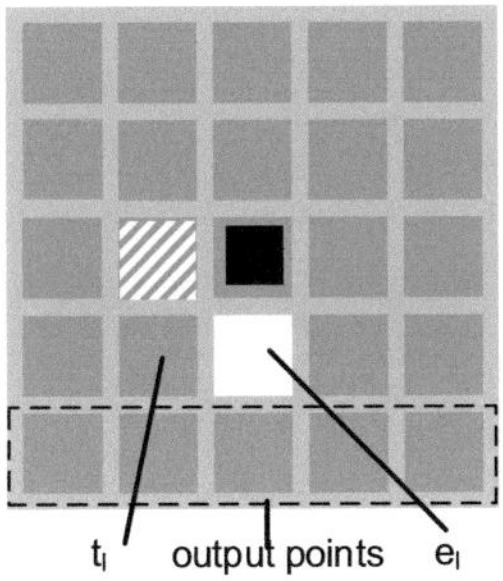

Figure 4.12: Case 3: escort e_l is not in the target cell t_l

When moving the escort towards the target cell, a detour might be necessary to avoid an involuntarily move of the assigned load (see Fig. 4.8). The decision rules for the target cell t_e of the escort are designed to avoid the necessity of a detour as far as possible.

To define t_e three phases are used. First, the direction to move e_l to has to be determined. Secondly, a check for a necessary detour has to be done. Thirdly, the according next AGV move has to be chosen.

Phase 1: Direction to Move Escort e_l to

To define the direction to move the escort e_l in, the horizontal and vertical distances between the assigned escort e_l and the target cell of the load t_l are considered. If the horizontal resp. vertical distance is bigger, the escort e_l is moved into the direction of the bigger difference, in order not to create the necessity of a later detour (see Fig. 4.13, left).

If the distances are equal, the escort e_l is moved parallel to the line of load l and target cell t_l to make sure that the assigned load is not unwillingly moved (see Fig. 4.13, right).

The target cell for the escort t_e is therefore selected to be the neighboring cell to the escort e_l in the determined direction.

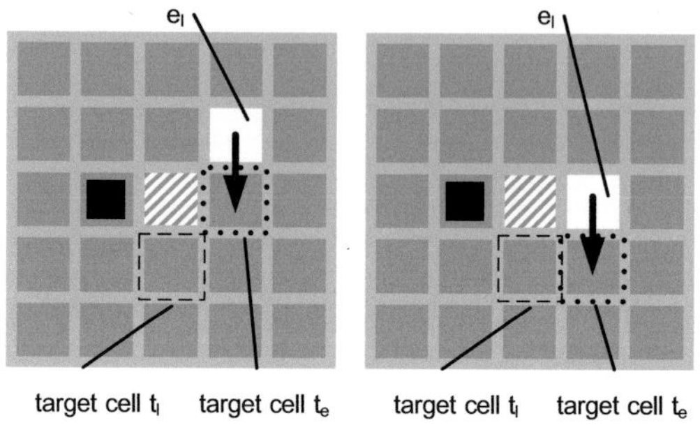

Figure 4.13: Phase 1: Direction to move the escort e_l to (left: vertical distance of e_l and t_l bigger, right: both distances equal)

Phase 2: Check for Detour

Based on the direction a check for a necessary detour is done. A detour will be necessary if the assigned load l, the assigned escort e_l, and the target cell for the load t_l are in one line and the escort is in the neighboring cell to the assigned load (see Fig. 4.8). In this case, the target cell for the escort t_e is redefined to the neighboring cell in orthogonal direction of the line of l, e_l, t_l. As there are two alternatives, t_e is chosen according to the general assumptions (see Section 4.1 and Appendix A).

Phase 3: Define Next AGV Move

After the next target cell for the escort t_e is defined, the according next single AGV move is chosen. If the AGV is in t_e, the load in this cell can be moved to the escort position and thus move the escort to t_e (see Fig. 4.14, left). If the AGV is not yet in t_e, it is moved towards t_e (see Fig. 4.14, right).

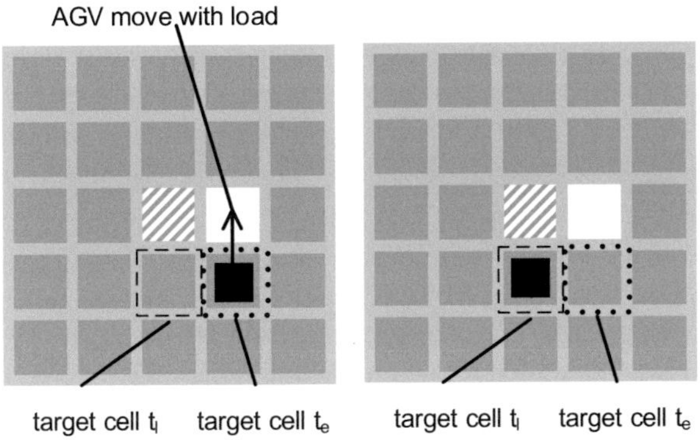

Figure 4.14: Phase 3: AGV move to transport the escort to its target cell t_e (left: AGV load to be moved to t_e, right: AGV not under this load)

4.6 Step 4: Handling of Deadlocks and Livelocks

As all AGVs determine their next single move individually, there will be situations of conflicts between AGVs blocking each other eventually leading to deadlocks. A deadlock is a situation in which AGVs come to a stop and at least one AGV can never move on. Additionally, multiple AGVs might influence each other, leading to circling movements which are never exited (livelock). Therefore, a method to prevent or solve those situations has to be implemented to ensure system functionality at all times.

4.6.1 Deadlocks

A deadlock occurs in a situation of two or more vehicles blocking each other, i.e. each preventing the desired move of the other (see Fig. 4.15). Mayer (2009) describes an algorithm to prevent deadlocks in networks of conveying units (FlexConveyors) which are linked to each other (see Section 2.1). As these conveying units are linked to all their neighbors they can send and receive messages and, thus, get information about the usage and reservation of desired paths. Livelocks are prevented by avoiding specified system layouts.

In the GridFlow system with AGVs only the AGVs but not the cells of the grid are equipped to send or receive messages. To equip the cells with such technology would required additional infrastructure and investment. This would revoke the advantages of the GridFlow system with AGVs. Therefore the methodology of Mayer (2009) cannot be implemented in the proposed system design.

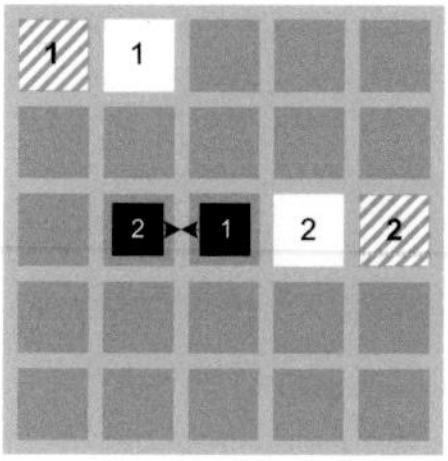

Figure 4.15: Situation of a deadlock

A similar methodology to prevent deadlocks and livelocks before they occur, would require a centralized computer providing and updating the current as well as future system states according to reservations on cells. For a decentralized solution, the AGVs would have to do transaction handling. This would lead to very complex considerations and to very heavy data traffic as not only the current state would have to be considered but also reservations in the future. Additionally, the parallel decisions on reservations could easily lead to non-robust preplanning of paths.

Therefore, we did not include a deadlock or livelock prevention in the strategy LivePath. The strategy includes rules to solve deadlocks once they occur rather than trying to prevent them. To enable those rules, the ranks of the AGVs are needed (see Section 4.1). Each AGV is assigned to an individual rank with 1 being the highest rank. To resolve deadlocks, AGVs may order others with a lower rank to wait or retreat.

If an AGV has to wait because the cell it wants to move to, is blocked because of a move by another AGV, it will place a wait request. This means that it asks all other AGVs with a lower rank to wait. All AGVs with a lower rank will finish their active moves but afterwards will not move on while the wait request of the initiating AGV is active. The wait request will be cleared after the initiating AGV was able to move as desired.

Additionally, an AGV may make another AGV with a lower rank retreat by passing a back up request to the other AGV. An AGV will place a back up request if the determined move is not feasible because another AGV with a lower rank is blocking the cell, the initialising AGV wants to move to. In the example in Figure 4.15, AGV 1 will tell AGV 2 to clear the cell as AGV 1 wants to move there.

The AGV that receives the back up request will always finish its active move before backing up. After the move is finished, it will dismiss its determined next move if it still has to back up. It will then look for the closest cell which is not occupied by another AGV according to the rectangular metric. The neighboring cell in the direction of the closest cell without AGV is defined as its next target cell t_b for the back up move. If t_b is not occupied by an AGV, the AGV backing up will move to t_b and the initiating AGV can continue. If t_b is occupied by another AGV, this AGV also has to back up if its rank is lower than the one of the initial AGV (see Fig. 4.16). Therefore, an AGV backing up with its target cell t_b being occupied, will ask the AGV in t_b to back up with the rank of the initiating AGV. This means that the initial rank is passed on, and all AGVs with a lower rank than the initiating AGV will also back up if this is required. In the example in Figure 4.16, AGV 2 will back up for AGV 3 although its rank is higher, because the initial back up request was issued by AGV 1 and AGV 3 passes on the rank 1.

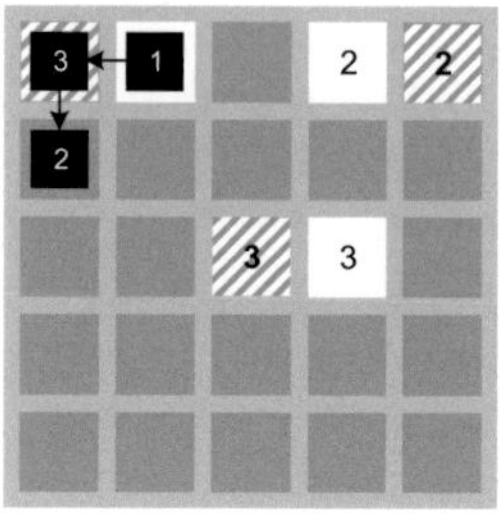

Figure 4.16: AGVs 3 backing up for AGV 1, forcing AGV 2 to back up as well

4.6.2 Livelocks

Additional to deadlocks, the system might experience livelocks. This means that at least one vehicle is moving in a recurring cycle without ever leaving this looping movement. The loop might involve any number of recurring moves. If this happens, the assigned load will never be retrieved resp. replenished. To detect and solve livelocks, the movements of all vehicles are tracked and analysed for looping movements. A looping movement may in some cases not result in a livelock but might be a temporary phenomenon. But to ensure system functionality, the livelock handling is started whenever a looping movement is detected.

For the loop detection all moves of the vehicle while performing the active replenishment or retrieval (vehicle is assigned to the active load) are considered:

- Loaded loop: AGV moves the same load between the same two cells in the opposite direction (movement there and back) or in the same direction (movement in a cycle)
- Unloaded loop: AGV moves between the same two cells in the opposite or same direction while the assigned load is still in the same position as it was when the first move between the two cells occurred

If an AGV is looping it will place a wait request. This means that AGVs with a lower rank will be held by the wait request while the initiating AGV is in a looping movement. The initiating AGV will cancel the wait request when it was able to exit its looping movement.

However, an AGV in a looping movement might need to push another AGV back to leave the loop. Additionally, a waiting AGV might block another AGV with a higher rank than the rank of the wait request, which might lead to deadlock situations. In the example of Figure 4.17 AGV 2 placed a wait request holding AGV 3. This would lead to AGV 1 waiting and, thus, to a possible deadlock situation.

Therefore, there is one exception from the waiting rule. An AGV that is held by a wait request may move if it receives a back up request with the same or a higher initial rank than the rank of the wait request.

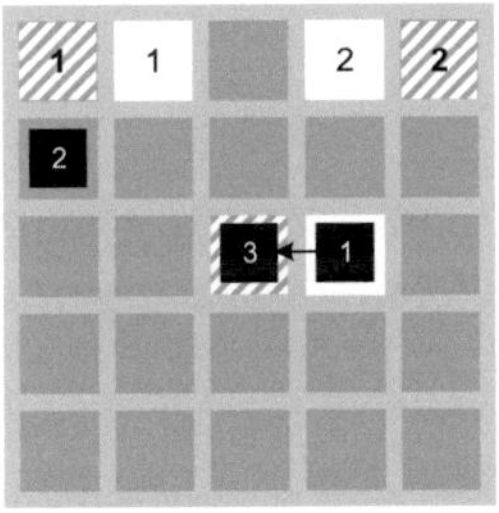

Figure 4.17: Example for a possible deadlock if AGV 2 issued a wait request

4.6.3 Simulation of LivePath System

To enable an assessment of solution quality as well as to give insights on system behaviour, the strategy LivePath was implemented in the simulation software AnyLogic (TM).

System Control

We used an agent-based model to model the AGVs. However, in this simulation model there was still a central unit timing the decision of the AGVs and providing shared information. The central unit clocked the decisions of the AGVs and made sure that the AGVs took their decisions in order of their ranks. This ensured the vehicle with the highest rank to take its decision first in each time step. Additionally, the central unit provided shared information, e.g. about the blocked cells to all AGVs. Information was therefore only stored in this instance, and validity was ensured at all times as all AGVs updated the central information sequentially due to the clocked decision taking.

Thus, the simulation model did not provide a completely decentralized implementation of the strategy. In a real world decentralized implementation, AGVs would take decisions simultaneously. Therefore, the AGVs would need to ensure the validity of their decisions by an additional transaction handling by communicating with the other AGVs. Furthermore, the central information would not be available in a totally decentralized system. Each AGV would store and update its information based on received messages. To ensure that the information is always up to date, additional messages and transaction handling might be required. Additionally, AGVs have to confirm the reception of messages to avoid decisions based on out of date information which might lead to collisions, deadlocks or livelocks.

However, the simulation model in this thesis enabled us to evaluate the functionality of the strategy sufficiently for this theoretical purpose. We tried to provide insights on system behaviour and opportunities of the strategy LivePath. We were not trying to prove the real world versatility of this strategy.

Simulation Mode

In the 'retrieval scenarios' mode, the simulation model considered single retrieval scenarios to compare the strategy results with the optimal solution (see Section 7). A given number of vehicles, requested loads, and escorts was analysed. A retrieval scenario started with a random or given set of positions and ended when all requested loads were taken from the grid.

In the 'continuous throughput' mode, the model considered replenishments. Whenever an AGV finished a retrieval, it left the grid with the load. To maintain a constant system density, after each retrieval a new replenishment was initialised. The waiting position of the initialised load at the grid edge was determined by a round robin procedure of all feasible waiting positions if nothing else is mentioned. The initialized load was placed into the waiting position with an AGV that moved the load into the grid.

Simulation Detail

The simulation model was also used to model different movement details.

The 'equal times' mode used equal movement times for all moves. This means that an AGV performed one move per time step independent of the move characteristics (e.g. loaded vs. unloaded move). We used this mode to give general insights on system behaviour (see Chapter 8 and Section 7).

The 'differing times' mode considered differing movement times. This means that the movement characteristics and their specific times for all necessary movement parts (e.g. steering or loading) were considered. Due to the clocked decision process all considered times were multiples of one second in this mode. This mode could provide more realistic results for a case with a specific technical implementation (see Nobbe (2015)).

In both simulation details a cell first had to be cleared completely before another AGV could start to move into this cell (see Chapter 6).

5 Deadlock and Livelock Behavior

To guarantee the functionality of the system, deadlocks and livelocks have to be prevented or resolved at all times. Mayer (2009) transfers deadlock conditions from computer operating systems to material flow systems. To avoid deadlocks at least one of the following conditions may not be fulfilled (Mayer 2009, p. 77f).

1. **Mutual exclusion:** processes require the exclusive use of resources
2. **Hold while waiting:** processes hold onto resources while waiting for additional required resources to become available
3. **No pre-emption:** processes holding resources determine when they are released
4. **Circular waiting:** closed chain of processes in which each process is waiting for a resource held by the next process in the chain

Only one vehicle can be in one cell at a time and each vehicle stays in its cell until it determines to leave the cell and is able to do so. Thus, the first three conditions are necessarily always fulfilled in the GridFlow system with AGVs and the strategy LivePath. Therefore, we have to make sure that the fourth condition is not true.

Additionally to deadlocks, livelocks have to be considered. AGVs might not be in a process of circular waiting but one AGVs decision might influence another AGVs decision and vise versa, leading to circular movements. Thus the vehicles are caught in a livelock. Therefore, circular waiting as well as circular movements have to be prevented or resolved. The strategy LivePath does not try to avoid deadlocks and livelocks but will resolve them on occurrence (see Section 4.6).

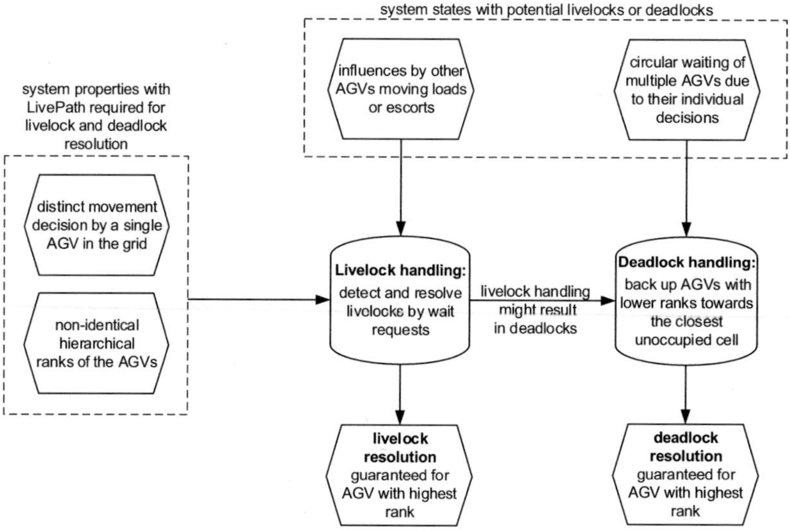

Figure 5.1: Concept of Livelock and Deadlock resolution

Figure 5.1 gives the concept of livelock and deadlock resolution with LivePath. The resolution is based on two system properties of LivePath:

- Distinct movement decision by a single AGV in the grid
- Non-identical hierarchical ranks of the AGVs (see Section 4.1)

The non-identical hierarchical ranks are described in the general system properties. We will show that the decision process of a single AGV in a grid always leads to a deterministic decision for the next move in Section 5.1. The decision has to be deterministic as random choices may lead to livelocks. If the vehicle for example chose one of two possible output points randomly, it might choose the other one in the next step and so on, leading to a cycle of changing decisions and thus a livelock.

Next, we will analyse system states with potential livelocks or deadlocks by analysing the influences of an additional AGV and we will describe how these are handled in the strategy LivePath. Livelocks may occur due to influences of other AGVs that move loads or escorts and deadlocks may occur due to circular waiting of multiple AGVs. These influences are further discussed in Section 5.2.

The system states with potential livelocks or deadlocks are handled in Step 4 of the strategy (see Section 4.6). Due to the livelock and deadlock handling at least the AGV with the highest rank (AGV1) will be able to take its requested load from the grid. If AGV1 is replenishing, all other AGVs enter the grid afterwards in replenishing mode, because replenishing AGVs have lower ranks (see Section 4.1) and the ranks are assigned in order of grid entrance. Therefore, AGV1 will eventually finish its replenishment and start a retrieval. After AGV1 left the grid the ranks will be updated and the AGV 2 will become AGV1. Thus, even if only AGV1 is able to perform its task, sequentially all AGVs will be able to do so and eventually leave the grid as they will be AGV1 eventually. Therefore, it is sufficient to ensure that the AGV with the highest rank is not caught in a deadlock or livelock. The ability of the livelock and deadlock handling to enable at least the AGV with the highest rank to perform its task is described in Sections 5.3 and 5.4.

5.1 Decision Diagram with a Single AGV in the Grid

The decision process of a single AGV with the rules of LivePath (see Chapter 4) is given in Appendix A. For each system state with the AGV being the only AGV in the grid a deterministic decision is chosen for all system states. There cannot be system states that do not lead to a decision as there is always a decision for each "else"-case.

5.2 Potential Influences of an Additional AGV

In Step 1 the AGV determines the next target cell for the assigned load. First, it determines the output resp. the input point for the load. In this decision no positions of other AGVs, loads or escorts are considered. Therefore, an additional AGV will not have any influence on this decision. Secondly, the target cell for the load is determined, which might be the direct target cell or might be based on a partial fast path. In this decision the positions of the escorts in the neighborhood of the load are considered. Therefore, an additional AGV might influence the decision in a future step by moving the assigned load or a load in the neighborhood. This might lead to multiple AGVs influencing each other and therefore to circular movements in a livelock.

In Step 2 the AGV determines the escort which is used to move the load to its target cell. In this decision the positions of the assigned load and the escorts in the grid are considered. As described above, this imposes the danger of a livelock as an additional AGV might change those positions influencing the decision in the next time periods of the initial AGV.

In Step 3 the AGV determines the next AGV move which might be a loaded or an unloaded move. In this decision, the positions of the assigned load and the escort are considered. Therefore, a livelock might arise in this step. Additionally, the AGV might not be able to perform the desired move due to other AGVs blocking the desired target cell. As the AGV will wait for the target cell to be cleared, a circular waiting of multiple AGVs and thus a deadlock might arise.

In summary there are three potential influences of an additional AGV leading to a livelock or deadlock:

1. Change of the assigned load position by another AGV → danger of circular movements (livelock)
2. Change of an escort position by another AGV → danger of circular movements (livelock)
3. Multiple AGV waiting for a cell to be cleared by another AGV → danger of circular waiting (deadlock)

5.3 Livelock Handling

5.3.1 Livelock Detection

A Livelock may occur due to two types of cycles an AGV can be caught in:

1. moving a load (its assigned load or another load) in a cycle
2. moving in a cycle without a load

These cycles could include multiple cells in round trips or oscillating moves. An example for loaded cycles might be an AGV moving the same load between the same two cells again and again, because another AGV moves it back to its first cell again and again. Another example could be two AGVs influencing each others decision on the escort to use, resulting in both AGVs moving in cycles.

The conditions to detect the two types of cycles are given in Section 4.6.2. To detect the cycles, the moves of the vehicles are tracked. If an AGV is able to successfully replenish or retrieve a load, it cannot be involved in a livelock or deadlock at this moment. Thus all movement of an AGV while assigned to the active load are tracked and all earlier movements while assigned to another load may be neglected.

The indicator for a loaded cycle is the vehicle moving the same load between the same two cells in the same or opposite direction.

The second cycle cannot include a cycle of the assigned load as it is an unloaded cycle. Thus, the cycle will occur while the assigned load is not moving. Therefore, the indicator for the second cycle is the vehicle moving between the same two cells (in the same or the opposite direction) without the assigned load being moved in the meantime.

These two indicators are sufficient but not necessary conditions for a livelock. Therefore, also temporal phenomenons might be detected and the livelock handling is started more often than necessary (see Section 4.6.2). This will result in a poorer throughput but it ensures system functionality.

5.3.2 Livelock Resolution

Whenever an indicator for a livelock is detected, the AGV will place a wait request to all other AGVs forcing those with a lower rank to wait (see Section 4.6.2). This ensures that livelocks will be resolved, as in the worst case all AGVs apart from the one with the highest rank will have to wait. This situation will result in the moves according to the decision rules for a single AGV.

However, forcing AGVs to wait might resolve a livelock but could result in a deadlock, if the AGV with the highest rank is waiting for an AGV to move, which is held by the wait request of this AGV (see Section 4.6.2). Those cases will be resolved by the deadlock handling.

5.4 Deadlock Handling

A deadlock will occur if multiple AGVs wait for each other to clear a cell in a waiting cycle. To resolve deadlocks, an AGV is able to force another AGV with a lower rank to clear a cell (see Section 4.6.1). To do so, it will send a back up request to the AGV blocking its path. Additionally it will send a wait request to all other AGVs with lower rank to resolve a possibly existing livelock.

However, a back up request always overrules a wait request with the same or a lower rank to enable AGVs to clear the path. The AGV backing off will determine the closest unoccupied cell and move towards this cell (see Section 4.6.1). If there are several unoccupied cells in the same distance, the cell to move to is chosen according to the general rules (see Section 4.1). If the closest unoccupied cell is not a neighboring cell to the AGV that has to clear the path, another AGV will first have to clear the path. Therefore, the AGV asked to back up will pass the back up request on, using the rank of the initial AGV (see Section 4.6.1).

A livelock due to back up moves cannot occur. If the AGV with the highest rank is placing a back up request, it will also issue a wait request. Therefore, only those AGVs that need to move to clear the path for the AGV with the highest rank may move. These AGVs will move in a sequential movement to the unoccupied cell closest to the AGV that first received the back up request.

Thus, the AGV with the highest rank will always be able to perform the desired move after if forced one or multiple other AGVs to back up as long as there is at least one unoccupied cell in the grid. This requirement has to be ensured in the initial system state. Afterwards, a new AGV with a replenishment load may only enter the grid after an AGV with a requested load has left the grid. Additionally, AGVs with replenishment loads have a lower rank than those AGVs with a requested load.

5.5 Assessment of Livelock and Deadlock Handling

By the above described rules, livelocks as well as deadlocks in the strategy LivePath can be resolved in all system states at least for the AGV with the highest rank. As this AGV will eventually leave the grid, each AGV will eventually become the AGV with the highest rank (see Section 5). Therefore, system functionality can therefore be ensured at all times.

However, in this thesis the implementation of the strategy LivePath is not completely decentralized (see Section 4.6.3). Therefore, in a real world implementation several conditions have to be ensured. The sharing of the shared information as well as the ranks have to be ensured, e.g. by confirmed messages. Additionally, decisions in LivePath have to be taken in the rank sequence. This might be implemented by action handling in real world implementations.

6 Optimal Movements in GridFlow Systems with AGVs

To assess the solution quality of the strategy LivePath a method to calculate the optimal movements is required. Rohit et al. (2010) propose an integer program to minimize the number of moves to retrieve a single load to a fixed I/O point from a multiple escort GridFlow system with conveyors. Each time period represents one load movement from one cell to the neighboring cell, and movement is enabled by conveyors. The model is based on an integer model for the Rush hour game.

We propose an integer model to retrieve one or multiple loads by one or multiple AGVs to one or multiple I/O points in a multiple escort GridFlow system with AGVs. Each time period stands for one vehicle move to a neighboring cell. However, the model is not an extension of Rohit et al. (2010) but based on network flow problems as the inclusion of vehicle movements seemed to be easier with this approach.

6.1 Optimization Model

The linear program minimizes the duration until all requested loads stated in the initial state have left the grid. There are no replenishments or load that are only requested later considered. In this linear program it is assumed that requested loads leave the grid with one AGV, and AGVs will not reenter the system as replenishments are not considered. Therefore, the number of AGVs has to be equal or bigger than the number of requested loads. Additionally, a cell may not be occupied by two different AGVs in two consecutive time periods, i.e. a cell has to be cleared completely before another AGV may start to move into the cell.

This is implemented due to considerations of technical implementations ((Nobbe 2015)).

6.1.1 Notation

The following sets, parameters and decision variables were used.

Sets

- Cells in Grid $\mathbb{I} : i, j = 1, 2, 3, .., N + 1$
- Cell N+1 is sink for loads leaving the grid
- AGVs $\mathbb{V} : v = 1, 2, 3, .., V < N + 1$
- Loads $\mathbb{L} : l = 1, 2, 3, .., L < N + 1$
- Time periods $\mathbb{T} : t = 1, 2, 3, .., T$

Parameters

- Adjacency matrix $A(i,j) = \begin{cases} 1, & \text{move from } i \text{ to } j \text{ is possible} \\ 0, & \text{else} \end{cases}$
- Matrix of requested loads
  $R(l) = \begin{cases} 1, & \text{load } l \text{ is requested} \\ 0, & \text{else} \end{cases}$
- Matrix of initial occupancies of cells by vehicles
  $O(v,i) = \begin{cases} 1, & \text{vehicle } v \text{ occupies cell } i \\ 0, & \text{else} \end{cases}$
- Matrix of initial occupancies of cells by loads
  $O(l,i) = \begin{cases} 1, & \text{load } l \text{ occupies cell } i \\ 0, & \text{else} \end{cases}$

Decision Variables

- m_{vijt}: Move by vehicle v from i to j in t
- o^v_{ivt}: Occupancy of cell i by vehicle v at the end of period t
- o^l_{ilt}: Occupancy of cell i by load l at the end of period t

6.1.2 Integer Programm

$$Min \quad \sum_t \sum_{l|R(l)=1} t \cdot o^l_{l,N+1,t} \tag{6.1}$$

$$s.t.$$

$$\sum_v o^v_{v,i,t} \leq 1 \ \forall t, i = 1,..,N \tag{6.2}$$

$$\sum_l o^l_{l,i,t} \leq 1 \ \forall t, i = 1,..,N \tag{6.3}$$

$$\sum_i o^v_{v,i,t} \leq 1 \ \forall v, t \tag{6.4}$$

$$\sum_i o^l_{l,i,t} = 1 \ \forall l, t | R(l) = 0 \tag{6.5}$$

$$\sum_i o^l_{l,i,t} \leq 1 \ \forall l, t | R(l) = 1 \tag{6.6}$$

$$\sum_t o^l_{l,N+1,t} = 1 \ \forall l | R(l) = 1 \tag{6.7}$$

$$o^v_{v,i,t} \leq o^v_{v,i,t-1} + \sum_{j|A(j,i)=1} o^v_{v,j,t-1} \ \forall v, i, t | t > 0 \tag{6.8}$$

$$o^l_{l,i,t} \leq o^l_{l,i,t-1} + \sum_{j|A(j,i)=1} o^l_{l,j,t-1} \ \forall l, i, t | t > 0 \tag{6.9}$$

$$o^v_{v,i,t} + \sum_{v' \neq v} o^v_{v',i,t-1} \leq 1 \ \forall v, i, t | t > 0 \tag{6.10}$$

$$\sum_{i,j,v,t|A(i,j)=0} m_{v,i,j,t} = 0 \tag{6.11}$$

$$m_{v,i,j,t} \leq \frac{o^v_{v,i,t-1} + o^v_{v,j,t}}{2} \quad \forall v, i, j, t | (t > 0, A(i,j) = 1) \tag{6.12}$$

$$o^l_{l,i,t-1} + o^l_{l,j,t} - 1 \leq \sum_v m_{v,i,j,t} \quad \forall l, i, j, t | (t > 0, A(i,j) = 1) \tag{6.13}$$

$$A(N+1, j) = 0 \ \forall j = 1,..,N+1 \tag{6.14}$$

$$m_{v,i,j,t=0} = 0 \ \forall v, i, j \tag{6.15}$$

$$o^v_{v,i,t=0} = O(v,i) \ \forall v, i \tag{6.16}$$

$$o^l_{l,i,t=0} = O(l,i) \ \forall l, i \tag{6.17}$$

$$m_{v,i,j,t}, o^v_{v,i,t}, o^l_{l,i,t} \in \{0,1\} \ \forall v, l, i, j, t \tag{6.18}$$

6.1.3 Explanation of Objective Function and Constraints

- 6.1: Minimize the number of time steps until all requested loads have left the grid.
- 6.2: There can only be one vehicle in a cell at a time.
- 6.3: There can only be one load in a cell at a time.
- 6.4: A vehicle can only be in one cell at a time and vehicles may leave the grid with a requested load.
- 6.5: A not requested load can only be in one cell at a time and may not leave the grid.
- 6.6: A requested load can only be in one cell at a time and requested loads may leave the grid at the end of their retrieval.
- 6.7: Each requested load must reach one of its feasible output points.
- 6.8: A vehicle can only occupy a cell if it occupied this cell or one of its neighbors before.
- 6.9: A load can only occupy a cell if it occupied this cell or one of its neighbors before.
- 6.10: There can either be a vehicle move into a location or a vehicle move out of the location, but not both in the same time step resp. a cell may not be occupied by different vehicles in consecutive time periods.
- 6.11: Vehicle moves can only occur between neighboring cells.
- 6.12: There can only be a move of vehicle v between two nodes if the vehicle occupies the nodes i and j in two consecutive time steps.
- 6.13: A load move requires a corresponding vehicle move.
- 6.14: Vehicles or loads may not reenter the grid.
- 6.15: In time step 0 there are no vehicle moves.
- 6.16: The occupancies by vehicles must reflect the initial setting.
- 6.17: The occupancies by loads must reflect the initial setting.
- 6.18: All variables are binary.

6.2 Tree Algorithm

As an alternative to the proposed binary program, the optimization might be conducted by a tree algorithm. Starting from the initial state all possible succeeding states can be derived, and a tree of all successors can be build. In each time step, each vehicle has 9 movement possibilities: no move, loaded move in one of the four directions or unloaded move in one of the four directions. If there is one AGV in the system, each state has 9 potential successors, if there are two AGVs each state has 81 potential successors, etc. A depth first search might be used to find the shortest path in the tree from the initial state to the retrieval of all requested loads. An initial bound has to be applied because otherwise the depth first algorithm might not terminate if the first considered path does not lead to a retrieval state but e.g. a movement loop.

6.3 Number of System States and Effects on Solution Time

The proposed integer program can be solved by classical branch-and-bound algorithms to obtain the optimal solution although it is NP-hard. The proposed tree shows exponential growth. The number of possible system states underlines this. The number depends on:

- Grid size = number of cells N
- Number of vehicles V
- Number of escorts E
- Number of requested loads R

6.3.1 Shared I/O Points

If all requested loads share the same I/O points the number of system states S_{shared} can be calculated by:

$$S_{shared} = \binom{N}{V} \cdot \binom{N}{E} \cdot \binom{N-E}{R} \tag{6.19}$$

The AGVs and escorts can be in any cell in the grid. The requested loads can be in any of the occupied cells ($N - E$). It does not matter which of the AGVs, escorts or load is in what cell as each escort can be used by each AGV, each requested load can be moved by each AGV, and the requested loads all have the same I/O points. The number of possibilities can therefore be calculated by k-combinations.

6.3.2 Individual I/O Points

If the requested loads have individual I/O points differing from the I/O points of the other requested loads, the number of system states S_{ind} can be calculated by:

$$S_{ind} = \binom{N}{V} \cdot \binom{N}{E} \cdot \frac{(N-E)!}{(N-E-R)!} \tag{6.20}$$

In this case it does matter which requested load is in what cell. The number of possibilities to place the requested loads can therefore be calculated by a k-permutation.

6.3.3 Influence on Solution Times

Table 6.1 gives examples of the number of system states for different configurations.

The NP-hardness of the programm and the exponential growth of the tree result in expected solution times that prohibit the implementation of a central controller determining the optimal movements at all times. A heuristic or determined strategy is required for real world application.

	number of system states	
configuration	individual I/Os	shared I/Os
5x5, 1 AGV, 1 escort, 1 request	15,000	15,000
5x5, 1 AGV, 1 escort, 2 requests	345,000	172,500
5x5, 1 AGV, 1 escort, 3 requests	7,590,000	1,265,000
5x5, 1 AGV, 1 escort, 4 requests	159,390,000	6,641,250
5x5, 5 AGVs, 5 escorts, 5 requests	$5.25176E + 15$	$4.37646E + 15$

Table 6.1: Examples of numbers of system states

Furthermore, a centralized control strategy holds many drawbacks especially in regard to system flexibility. As flexibility is a main opportunity of GridFlow systems with AGVs, we propose the decentralized control strategy LivePath for GridFlow systems with multiple AGVs.

7 Assessment of LivePath Solution Quality

To assess the solution quality of LivePath we optimized and simulated different configurations of GridFlow systems with AGVs. For each configuration 1000 retrieval scenarios were analysed with the tree algorithm and depth first search. The initial bound was the retrieval time of LivePath, as the simulated retrieval time may not be shorter than the optimal one.

Due to the long solution times of the optimization, only small configurations with only 1 AGV and up to 5 escorts could be analysed. No configuration with 2 AGVs offering useful insights could be solved optimally. Very small configurations with 2 AGVs would not present representative data, as the rules of the strategy LivePath will not unfold their potential in small configurations with 2 AGVs.

7.1 Influence of Grid Size and the Number of Escorts

7.1.1 Distribution of Differences

Figure 7.1 shows the histogram of the absolute difference in retrieval time between optimization and LivePath for a 3x3 configuration and 1 AGV. With 1 escort in 96.5% of all retrieval scenarios, there was no difference, and LivePath resulted in the same retrieval time as the optimization. This percentage decreased with increasing number of escorts. Presumably this is due to the increasing number of movement opportunities with more escorts in the system.

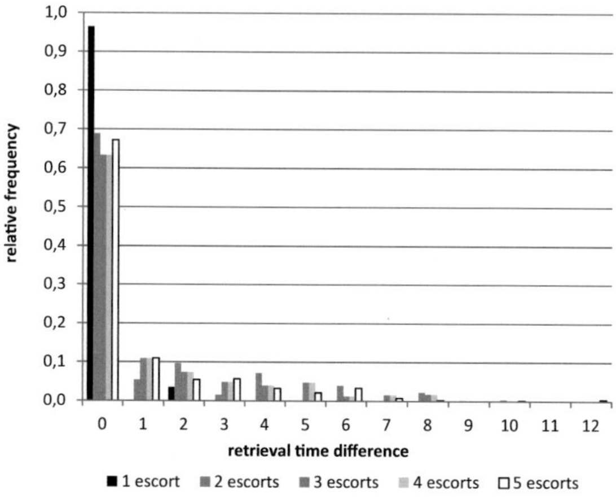

Figure 7.1: Histogram of absolute difference in retrieval time for a 3x3 configuration with 1 AGV

7.1.2 Average Absolute Differences

Figure 7.2 shows the average absolute differences in retrieval time for different grid sizes and numbers of escorts. Bigger grids showed bigger absolute differences. This suggests the solution quality to decrease with grid size.

A maximum of the absolute difference in dependence of the number of escorts could be obtained from the data. The corresponding number of escort increases with increasing grid size. This suggests the average differences to increase with the number of escorts, reaches a maximum and decrease slightly for a high number of escorts. The decrease for high numbers of escorts however seems to be smaller than the increase for smaller number of escorts. The data from the scenarios that could be solved optimally is not sufficient to derive a mathematical function.

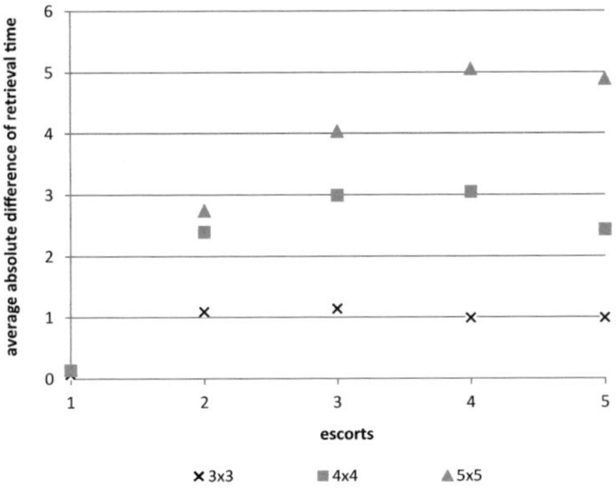

Figure 7.2: Average absolute retrieval time differences with 1 AGV

7.1.3 Average Relative Differences

The relative difference is calculated by dividing the absolute difference of the retrieval times by the optimal retrieval time.

The average relative differences showed the same behavior than the average absolute differences. Although there are some differing values, overall the average relative difference can be considered to be bigger for bigger grids.

For 3x3 and 4x4 grids the average relative difference also showed a maximum difference in dependence of the number of escorts. The 5x5 grid did not show this behaviour for the analysed numbers of escorts. A decline for bigger numbers of escorts can be assumed but not verified with this data.

7.2 Assessment of Solution Quality

The analysis for a single AGV suggests the solution quality of LivePath to decreases with increasing grid sizes. The data suggests a maximum value

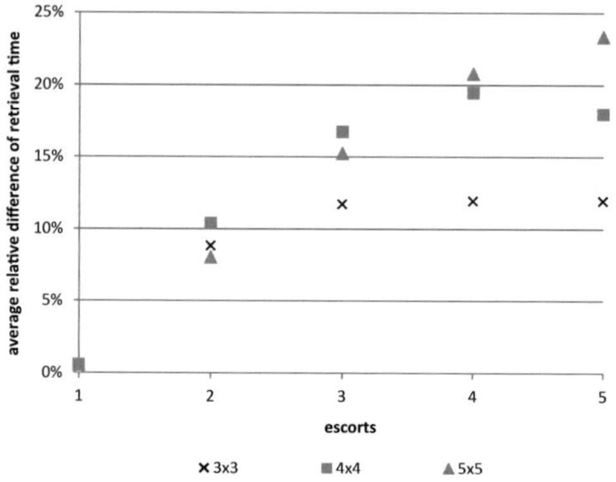

Figure 7.3: Average relative retrieval time differences with 1 AGV

of average absolute and relative difference in dependence of the number of escorts. Results for multiple AGV could not be calculated due to the solution times of the Optimization. The behavior of the solution quality with multiple AGVs therefore could not be assessed.

Figure 7.4 shows the average relative time differences for all considered configurations. The maximum average relative difference is 23% for a 5x5 grid with 5 escorts. This means that the strategy LivePath requires 123% of the optimal retrieval time.

Therefore, the strategy LivePath shows promising results, but also further potential, concerning the solution quality for a single AGV. However, as the assessment of multiple AGVs is not possible, the significance of the assessment of the solution quality is limited.

Still, the strategy LivePath offers promising results. Therefore the system behavior with LivePath is analysed further.

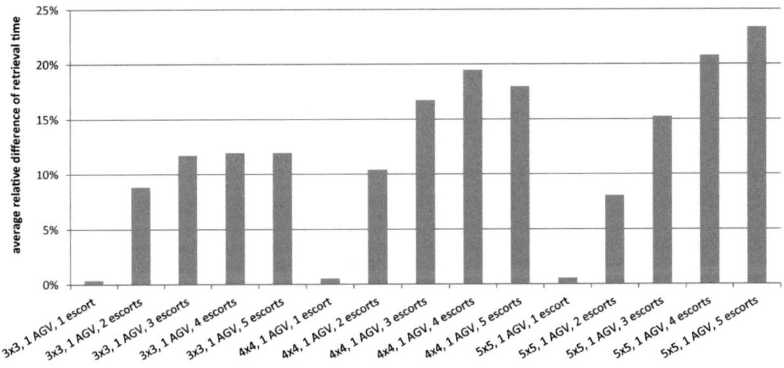

Figure 7.4: Average relative time differences

8 System Behaviour with LivePath

In order to design real world implementations of GridFlow systems with AGVs and the strategy LivePath, knowledge about the system behavior in dependence of the system parameters has to be available.

We wanted to derive insights on the system behavior of GridFlow systems with AGVs and the strategy LivePath. Therefore, we considered the remaining system parameters apart from the movement strategy. To derive general insights, we used the simulation detail of equal movement times (one move per time period independent from the move characteristics (see Section 4.6.3)), because we did not want the analysis to be influenced by vehicle parameters. We analysed this influence separately (see Chapter 9).

To ensure the validity of the simulation results, the method of analysis and result verification (see Section 8.1) as well as the used indicators (see Section 8.2) are described first.

Afterwards the system behavior is analysed. System performance can the influenced by the following system parameters (see Section 3.2):

- size of the grid
- layout of the grid
- number of escorts
- number of vehicles
- Number and positions of I/O-points
- movement strategy for retrieval orders and replenishments

8.1 Analysis Method and Verification of Simulation Results

An equal number of AGVs and assigned loads was used in all configurations as vehicles do not cooperate and only handle one request at a time. Idle times of vehicles or waiting times of requested loads for a free vehicle were therefore not considered, as we wanted to derive insights on the maximum throughput. To maintain a constant system density a replenishment was initialized after each retrieval (see Section 4.6.3).

In the starting state of each simulation run, all vehicles retrieved a requested load. There were no replenishments in the starting state. A replenishment was initialized after a vehicle finished the retrieval of a requested load. The starting state was determined by choosing the positions of the vehicles, requested loads, and escorts randomly in the grid.

To derive reliable results from simulation, the analysis should be based on data from a steady state. Furthermore, the structural influence of the random number generator has to be analysed by doing multiple runs using different seed values (Wenzel et al. 2006).

8.1.1 Steady State Results

In each simulation run 10,000 retrievals were simulated, and the first 1,000 were cut from the consideration. Therefore, the analysis is based on the last 9,000 retrievals. To ensure that the cut of the first 1,000 retrievals is sufficient, we analysed the variation of throughput. The indicator we used is the in-run variability of the throughput per configuration. In each time step t of each simulation run the current throughput $tp(t)$ was calculated (see Section 8.2). For each simulation run the in-run variability $c_{ir}^2(tp)$ of the throughput $tp(t)$ of the last 9,000 runs was calculated. As 10 simulations runs were performed for each system configuration, we used the maximum in-run variability $\hat{c}_{ir}^2(tp)$ out of the 10 simulation runs to assess the result stability.

Figure 8.1 gives the frequencies of the maximum in-run variability $\hat{c}_{ir}^2(tp)$ for all considered system configurations. Most variabilities were smaller

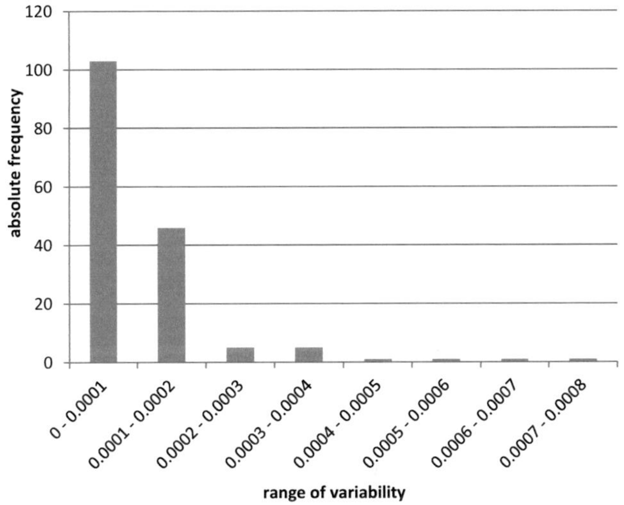

Figure 8.1: Frequencies of maximum in-run variabilities $\hat{c}_{ir}^2(tp)$ of throughput for different system configurations

or equal to 0.0001, and the maximum was 0.00079. Thus, the simulation of 10,000 retrievals and cut of the first 1,000 retrievals was considered to give steady state results.

8.1.2 Influence of Seed Values

For each configuration we simulated 10 runs with different seed values to analyse the influence of the random number generator and the starting state on system performance. For each run the average throughput $\overline{tp}$ (see Section 8.2) was calculated. To analyse the deviations between different runs for the same configuration we calculated the variability $c_{br}^2(\overline{tp})$ of the average throughput $\overline{tp}$.

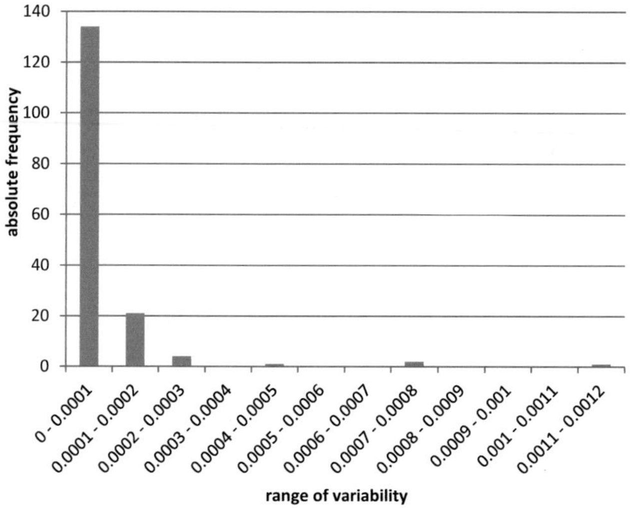

Figure 8.2: Frequencies of variability $c_{br}^2(\overline{tp})$ of average throughput between runs for different system configurations

Figure 8.2 gives the frequencies of the variabilities $c_{br}^2(\overline{tp})$ of the average throughput between runs for all considered system configurations.

Most variabilities were smaller or equal to 0.0001, and the maximum was 0.0012. Thus, the influence of seed values was considered to be minor. Thus, the calculation of the average values over the 10 runs for all used indicators (see Section 8.2) was considered to result in valid and representative data.

8.2 Indicators for System Analysis

To assess the system behavior, average values as well as fluctuations of different indicators for system behaviour have to be considered.

8.2.1 Average Values and Fluctuations

Average values give information about the behavior of the system in average and calculated average values can be used to estimate mean values of the system.

To gain insights on variations in the system, the variability can be used. The variability gives the relation of mean value $E(a)$ and standard deviation for any indicator a. We use the empirical standard deviation $s_e(a)$ as the estimations for mean value and standard deviation are calculated from experimental data (Wenzel, Rabe and Spieckermann 2006). The variability $c^2(a)$ in this thesis is calculated by:

$$c^2(a) = \frac{s_e^2(a)}{E(a)^2} \tag{8.1}$$

To derive insights on system behavior different indicators to assess the performance of the system may be used.

8.2.2 Indicators to analyse the System Behavior

Analysis showed that the simulation results gave steady state results, and the influence of the seed values and starting states can be neglected (see Section 8.1). Therefore, all indicators described in this section were calculated by using the average value between the values of the individual simulation runs.

Throughput

Throughput is given by the number of retrievals per 1000 time periods to avoid the visualization of small numbers. The current throughput $tp(t)$ in each time period t can be calculated using the cumulated number of retrievals $C_r(t)$ until t:

$$tp(t) = \frac{t}{C_r(t)} \cdot 1000 \tag{8.2}$$

The average throughput $\overline{tp}$ is calculated by:

$$\overline{tp} = \sum_t tp(t) \tag{8.3}$$

Proportion of Requested Loads in all Assigned Loads

In each configuration there is a constant number of AGVs with one assigned load each. In each time period an AGV can either be busy retrieving or replenishing. Therefore, the proportion of retrieving AGVs in relation to all AGVs (retrieving and replenishing) equals the proportion of requested loads in all assigned loads. If there are for example 5 AGVs in the grid and 3 are retrieving while 2 are replenishing, the proportion of requested loads in all assigned loads will be 0.8.

We use the proportion of requested loads in all assigned loads to gain insights on the utilization of the system. As replenishment loads only have to enter the grid and thus be transported from the waiting position outside the grid into the neighboring cell within the grid, the proportion of requested loads in all assigned loads should be big in a balanced system. If the proportion is small this can be used as an indicator for system imbalance and utilization problems as a lot of replenishment loads are trying to enter the grid.

The proportion of requested loads in relation to all assigned loads was tracked for the analysis whenever a requested load left the grid. Therefore, for a single AGV, the AGV was always retrieving when the proportion was tracked. Thus, the proportion for single AGVs will always be 100%.

Retrieval Time

The retrieval time gives the number of time periods it takes an individual requested load to leave the grid. In this thesis it is measured for each simulated retrieval. The retrieval time starts with the request of the load and the assignment of an AGV. It ends with the requested load and its assigned AGV leaving the grid. The first retrieval of an AGV during simulation starts with the AGV being in a randomly chosen cell. After the retrieval a replenishment will be initialized. When the replenishment enters the system, the AGV is assigned to the next request. Therefore, all later retrievals start with the AGV being in the top most row. Thus, the minimum retrieval time equals the number of grid rows.

Variability of Retrieval Time

The variability of the retrieval time is used to analyse differences in the retrieval time which might occur due to differing load positions or stochastic influences.

Absolute and Relative Waiting Time by the AGVs

The waiting time of the AGVs is considered because waiting will result in time loss and therefore in reduced throughput. A certain amount of waiting might not be evitable, even with optimal movements, but high proportions of waiting suggest an inefficient system as AGV spend most of their time waiting, instead of moving.

We gain the waiting time of an AGV by tracking the number of time periods an AGV is waiting for another AGV during the considered retrieval.

Additional to the absolute waiting time, we define the relative waiting time to be the proportion of time periods the AGV is waiting in relation to the total retrieval time for the assigned request. It is calculated by dividing the absolute waiting time by the retrieval time. This corresponds to the time share the AGV spends waiting.

Absolute and relative Number of Back Up Moves by AGVs

Additional to waiting, an AGV might be forced to clear a cell because an AGV with a higher rank wants to move there. The absolute number of back up moves is tracked for each retrieval. The relative number of back up moves can be calculated by dividing the absolute number of back up moves by the total number of AGV moves. It gives insight on AGV efficiency as it gives the proportion of undesired moves.

Connection of waiting time, back up moves, retrieval time and number of AGV moves

The total number of AGV moves per retrieval may be divided into back up moves and retrieval supporting moves. Additional to time periods the AGV is moving in, the AGV might be waiting for other AGVs. Thus, the total number of AGV moves plus the waiting time equals the retrieval time.

8.3 Influence of the Grid Size

To analyse the influence of the grid size on system behavior configurations with different grid sizes have to be analysed while the influence of all other parameters is limited as far as possible. Therefore, we defined system configurations that have different grid sizes, but an equal ratio of columns and rows as well as equal proportions of AGVs and escorts.

As only discrete numbers of columns, rows, AGVs ,and escorts are possible, the number of feasible configurations is limited. We analysed quadratic configurations, as this ratio of columns and rows offers many feasible grid sizes and an easy extrapolation. The analysed grid sizes vary from 9 to 256 cells (9, 16, 36, 64, 144, 225, 256 cells). In order not to base the analysis on a single relation of AGVs and escorts, we considered four different relations of AGVs and escorts.

8.3.1 Throughput

Figure 8.3 shows the average throughput for different grid sizes. There was a maximum throughput for all considered proportions of AGVs and escorts. The analysis therefore suggests that very small and very big grids show a smaller throughput than medium sized grids. The grid size with maximum throughput depended on the number of AGVs and escorts, and a general rule for the maximum throughput could not be derived from this data. As the analysis indicates a maximum throughput, it might be useful to subdivide bigger grids to multiple smaller ones to increase the throughput in real world implementations.

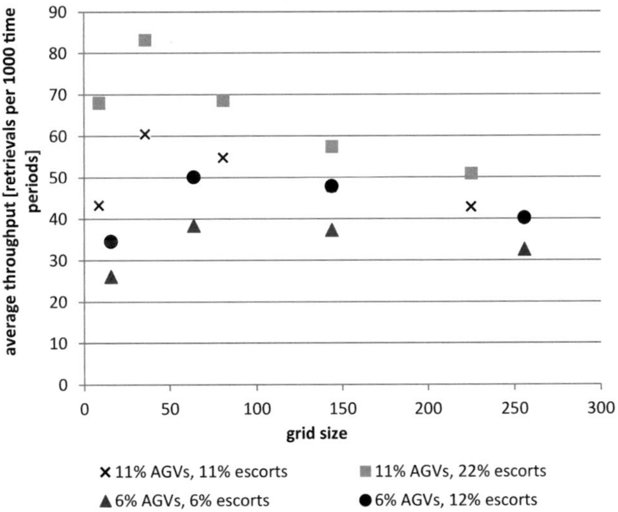

Figure 8.3: Average throughput for different grid sizes

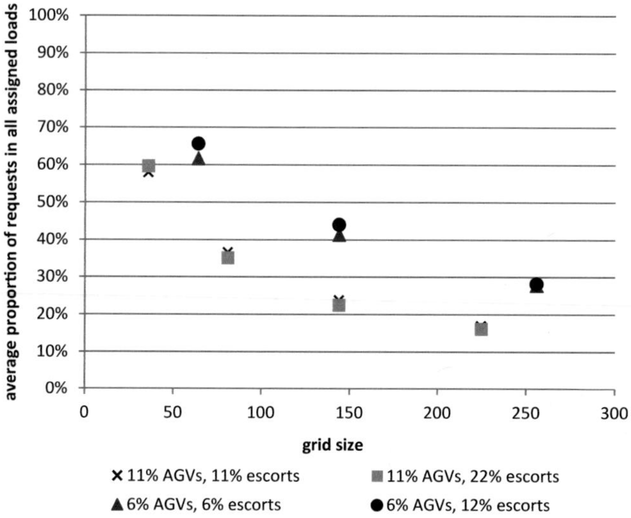

Figure 8.4: Average proportion of requested loads in all assigned loads for different grid sizes

8.3.2 Proportion of Requested Loads in all assigned Loads

Figure 8.4 gives the average proportion of requests in all assigned load. The proportion was decreasing for increasing grid sizes. This means that in bigger grids relatively more replenishments and fewer requests are active. A small proportion of active requests indicates a high utilization and an imbalanced system. The results corresponds with the decreasing throughput for big grids.

8.3.3 Retrieval Time

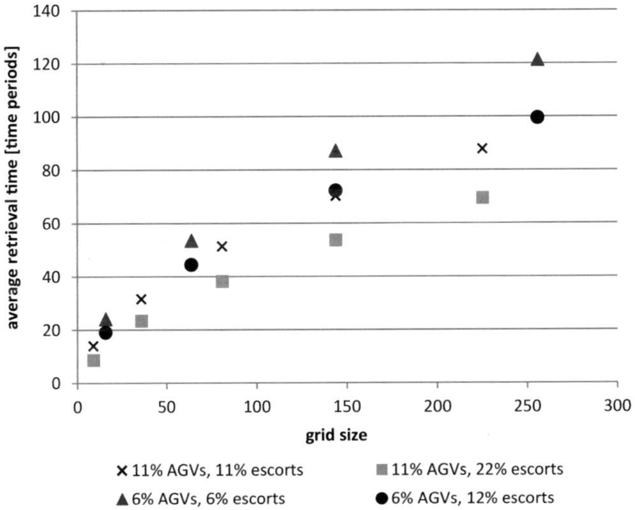

Figure 8.5: Average retrieval time for different grid sizes

The average retrieval times for all analysed configurations were increasing with increasing grid size (see Fig. 8.5). The data suggests a linear or polynomial incline. Table B.1 gives the result of the regression analysis of grid size and retrieval time. The regression analysis for the considered proportions of AGVs and escorts points to a polynomial growth of order 2. This suggests a decreasing slope of the retrieval time for bigger grids. However, bigger grids might show a different behavior. The relevance of bigger single grids for real world implementations with large loads, however, seems to be limited due to the behavior of the system throughput (see Section 8.3.1).

The average variabilities of the retrieval time varied between 0.3 and 0.5 with the grid size, indicating less than random variations. Supposedly, this is due to the AGV starting a retrieval in the top most row and the resulting maximal retrieval time. The average variability of the retrieval time was higher for small grids, and homogeneous for medium and big grids independent from the proportion of AGVs and escorts.

8.3.4 Waiting Time and Back Up Moves

The average absolute number of back up moves, the average number of AGV moves and the average retrieval time are displayed in one figure to visualize the proportion of waiting by the AGVs and of back up moves by the AGVs. As one time step equals one AGV move in the simulation mode, the proportion of waiting can be observed from the difference of retrieval time and number of AGV moves. The proportion of back up moves can be observed from the relation of the number of back up moves to the number of all AGV moves.

Figure 8.6 gives the average absolute number of back up moves, the average number of AGV moves and the average retrieval time for different grid sizes with 11% AGVs and 11% escorts. The other considered relations of AGVs and escorts showed the same structural behavior and, therefore, are not displayed here.

The data indicates that back up moves in average have a minor proportion in the total number of AGV moves. However, there were configurations with high proportions of back up moves, as the maximum percentage was 85%.

The average absolute waiting time is the difference of the average retrieval time and the average number of AGV moves. Figure 8.6 shows an increasing absolute waiting time for increasing grid sizes for the considered configuration. The relative waiting time also increased with increasing grid size, but with declining slope.

For the considered grid sizes and proportions of AGVs and escorts waiting time is not minor and may not be neglected. Further insights on waiting time in dependence of the number of AGVs and escorts is given in the Sections 8.5, 8.6 and 8.7.

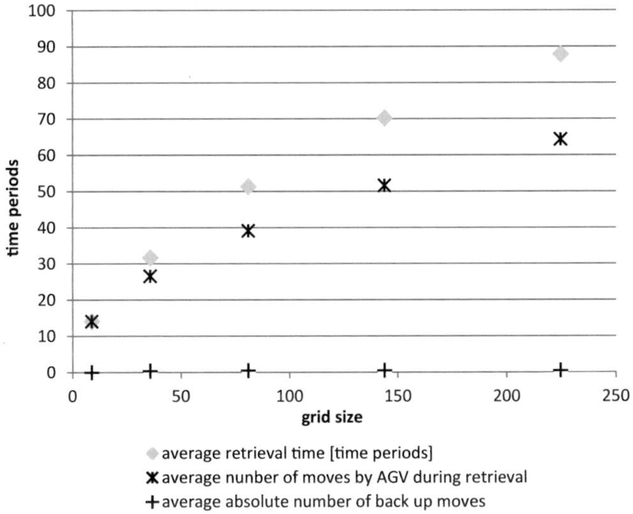

Figure 8.6: Average absolute number of back up moves, average number of AGV moves and average retrieval time for different grid sizes with 11% AGVs and 11% escorts

8.3.5 Summary

We analysed different grid sizes with quadratic layouts as well as with equal proportions of AGVs and escorts. The analysis suggests that there is a grid size providing a maximum throughput. The average proportion of requests in all assigned loads supports the hypothesis of decreasing throughput for bigger grids, as AGVs spend ever more time on replenishments, which points to a high utilization and an imbalanced system.

For small grids, the increase of throughput is supposedly due to the increasing absolute number of AGVs. Therefore, more requests may be handled at a time and the throughout increases. For bigger grids, however, the increasing waiting time, retrieval time, and decreasing proportion of requests in all assigned loads outweighs the effect of additional parallel requests. They even lead to an overloaded system with a lot of waiting replenishments and few requests.

The optimal grid size, however, seems to depend on the combination of the other system parameters and may not be obtained generally. Due to the analysis, it might be useful to subdivide big grids to multiple smaller grids in real world implementations.

8.4 Influence of the Grid Layout

To analyse the influence of the grid layout, we chose a grid capacity of 120 cells as this provides many feasible rectangular layout options. Considered layouts are denoted with XxY (X=number of columns, Y=number of rows, see Section 3.2). Narrow grids have a small number of columns in relation to the number of rows. Wide grids have a big number of columns in relation to the number of rows.

As we consider a system density of 90% to be feasible for a high density GridFlow system with AGVs, we used 12 escorts for our analysis. In order not to base the analysis on a single number of AGVs, 1, 6, and 12 AGVs were used.

8.4.1 Throughput

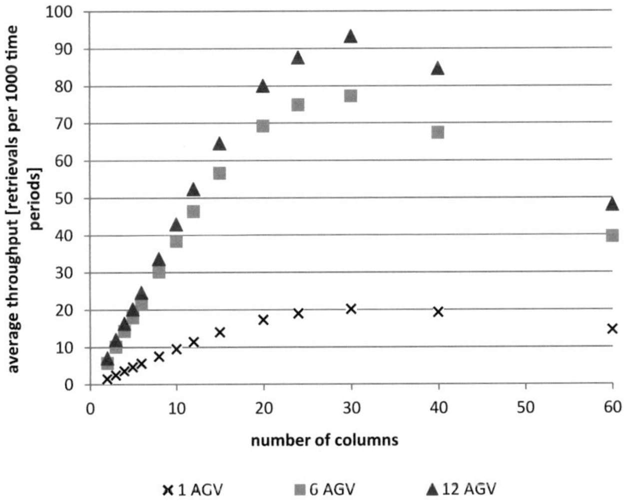

Figure 8.7: Average throughput for different grid layouts

Figure 8.7 gives the average throughput for different grid layouts. For all considered numbers of AGVs the layout with 30 columns and 4 rows resulted in the maximum throughput. Therefore, the data suggests an optimal ratio of columns and rows of $\frac{30}{4} = 7.5$ for a grid size of 120 cells. The analysis indicates a decreasing throughput if grids grow narrower. Wider grids also result in decreasing throughput, but the decrease is not as big as the decrease in throughput for narrower grids.

8.4.2 Proportion of Requested Loads in all assigned Loads

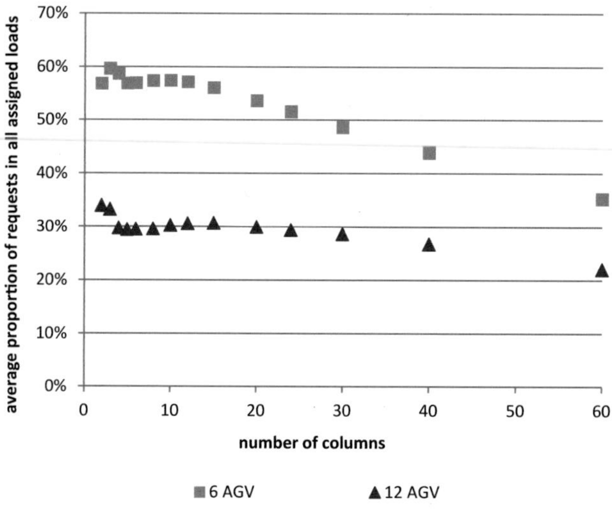

Figure 8.8: Average proportion of requested loads in all assigned loads for different grid layouts

Figure 8.8 gives the average proportion of requests in all assigned loads. For a single AGV no behaviour can be derived as only retrieval time periods are considered (see Section 8.3.2). For six and twelve AGVs, there was a decline in the proportion of requests for wide grids. This means that the AGVs spend more time with replenishments for wider grids than for narrow ones. For very narrow grids with 5 columns or less and 6 AGVs the proportions do not show an evident trend. However, the differences of 3% and the throughput behaviour of very narrow grids (see Section 8.4.1) do not call for closer attention.

The decreasing proportion of requests for wide grids indicates a high utilization and an imbalanced system. The results correspond with the decreasing throughput for wider grids (see Section 8.4.1).

8.4.3 Retrieval Time

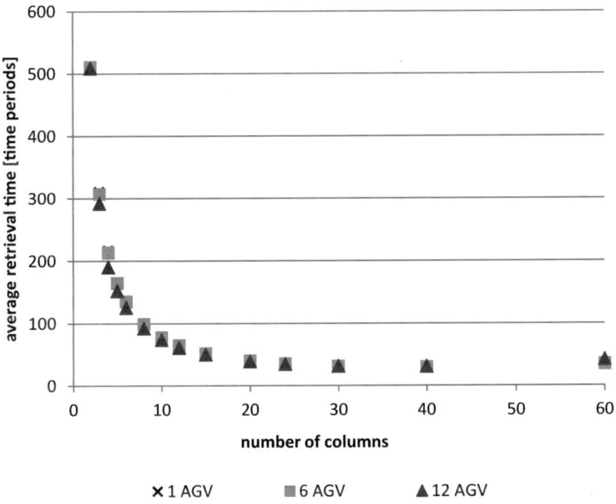

Figure 8.9: Average retrieval time for different grid layouts

Figure 8.9 gives the average retrieval time for different grid layouts. The retrieval times for all analysed configurations were decreasing with increasing numbers of columns. However, for very wide grids the retrieval time inclines again. The considered configurations showed the minimum retrieval time for a grid with 40 columns and three rows (see Fig. 8.5). Very wide grids do not seem to hold importance for real world implementations due to the throughput behaviour (see Section 8.4.1). Neglecting the data of the grid with 60 columns, the data suggests a hyperbolic decline. Table B.2 gives the result of the hyperbolic regression analysis of grid layout and retrieval time for the considered numbers of AGVs. The hyperbolic regression analysis gave a minimum coefficient of determination of 0.97 and thus supports the hypothesis of a hyperbolic decline for the considered grid layouts (neglecting data for 60 columns).

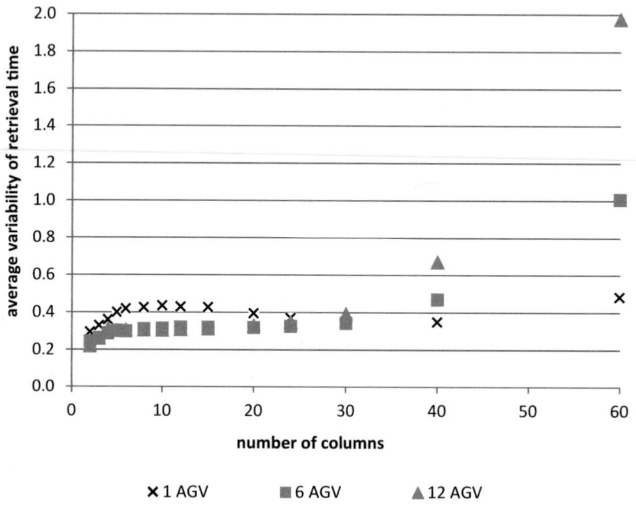

Figure 8.10: Average variability of retrieval times for different grid layouts

Figure 8.10 gives the average variability of the retrieval times for different grid layouts. We saw a strong incline in variability for very wide grids with 40 and 60 columns. For smaller numbers of columns the average variability showed a rather homogenous behaviour.

For very wide grids we saw big variabilities especially for six and twelve AGVs. For twelve AGVs the variability even indicates a worse than random behaviour. This behaviour supposedly is related to waiting time and back up moves of the AGV.

8.4.4 Waiting Time and Back Up Moves

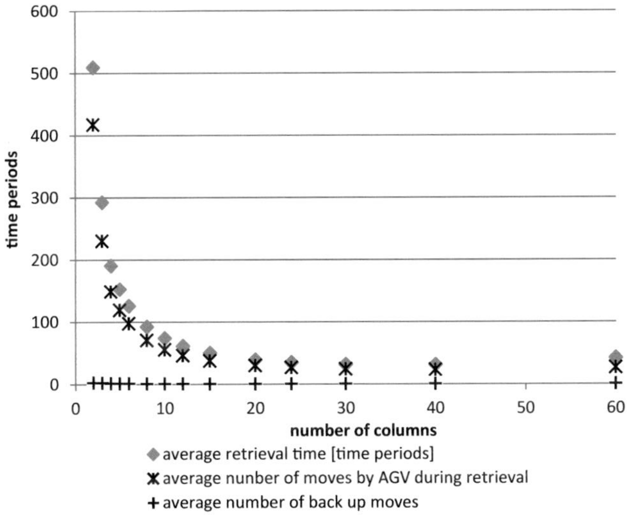

Figure 8.11: Average absolute number of back up moves, average number of AGV moves and average retrieval time for different grid layouts with 12 AGVs and 12 escorts

Figure 8.11 gives the average absolute number of back up moves, the average number of AGV moves in total and the average retrieval time for different grid layouts with 12 AGVs and 12 escorts. The other considered numbers of AGVs and escorts showed the same structural behavior and, therefore, are not displayed here.

The data indicates that back up moves in average had a minor proportion in the total number of AGV moves. However, there were configurations with high proportions, and the maximum proportion was 95%.

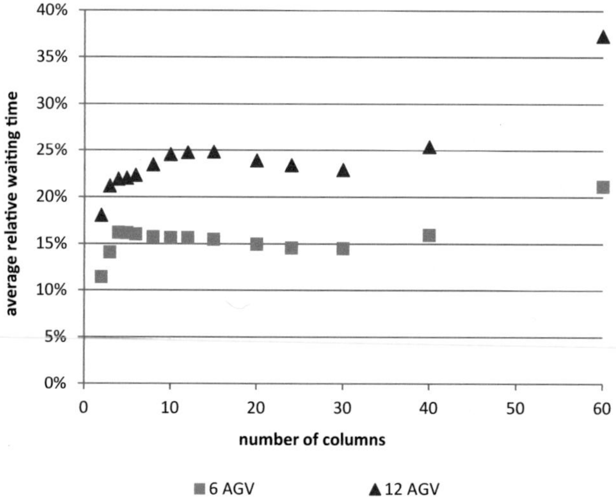

Figure 8.12: Average relative waiting time for different grid layouts

Figure 8.11 shows a decreasing absolute waiting time for increasing numbers of columns for the considered configuration. However for 40 and 60 columns the absolute waiting time increased. This might be due to very wide grids with very few rows offering a lot of blocking potential for vertical AGV moves.

The relative waiting time in the considered configuration increased for narrow grids, was homogenous for medium numbers of columns and increased again for grids with more than 30 columns (see Fig. 8.12). With the considered grid layouts and numbers of AGVs and escorts waiting time is not minor and may not be neglected. Further insights on waiting time in dependence of the number of AGVs and escorts is given in the Sections 8.5, 8.6 and 8.7.

8.4.5 Summary

We analysed 14 different rectangular grid layouts with a capacity of 120 cells, 12 escorts and three different numbers of AGVs. The analysis suggests an optimal ratio of columns and rows providing the maximum throughput. The average proportion of requests in all assigned loads supports this hypothesis of decreasing throughput for wider grids as AGVs spend ever more time on replenishments, which points to a high utilization and an imbalanced system. This is in accordance with the increasing relative waiting time for wider grids.

For narrow grids the increase of throughput with the number of columns is supposedly due to the increasing number of columns offering more movement options.

The optimal grid layout for 120 cells seems to be 7.5 but other grid capacities may not be implemented with this ratio and show an other optimal layout. Therefore, a general optimal ratio may not be obtained from this analysis. The data, however, points to the advantage of grids with more columns than rows.

In real world applications, the chosen layout might depend on other factors such as available space. Additionally, in grids with non-quadratic cells the optimal layout might differ, and differing movement times due to move characteristics might alter the optimal layout.

8.5 Influence of the Number of Escorts

To analyse the influence of the number of escorts, we chose a grid capacity of 120 cells with 30 columns and 4 rows, as this layout maximizes the throughput for this capacity (see Section 8.4). We varied the number of escorts for this layout with a constant number of AGVs. Analogous to Section 8.4 we considered 1, 6 and 12 AGVs.

8.5.1 Throughput

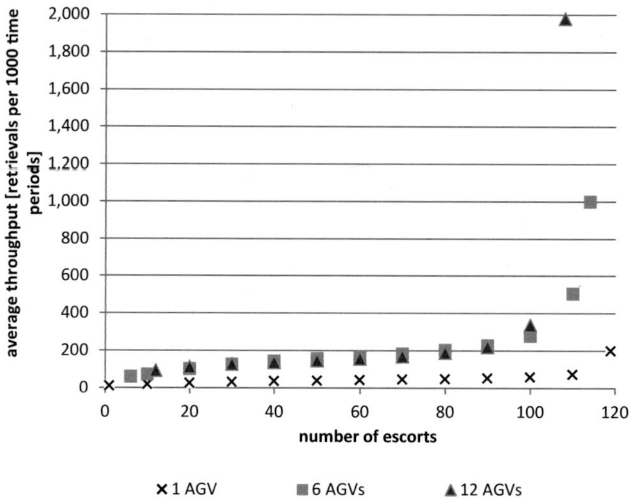

Figure 8.13: Average throughput for different numbers of escorts

Figure 8.13 gives the average throughput for different numbers of escorts. With all considered numbers of AGVs the throughput increased with the number of escorts. Therefore, high density grids provide a smaller throughput than grids with many escorts. For grids with a very low system density smaller than 16.7% (more than 100 escorts) the average throughput increased strongly.

The data from grids with up to 100 escorts suggests exponential growth for six and twelve AGVs. Regression analysis supports this. For 1 AGV linear growth seems to be more likely (see Table B.3).

8.5.2 Proportion of Requested Loads in all assigned Loads

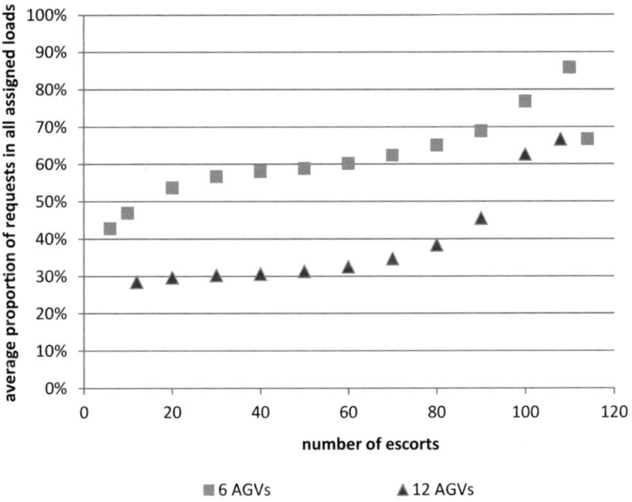

Figure 8.14: Average proportion of requested loads in all assigned loads for different numbers of escorts

Figure 8.14 gives the average proportion of requests in all assigned load. For a single AGV no behaviour can be derived as only retrieval time periods were considered (see Section 8.3.2). For six and twelve AGVs we saw an incline in the proportion of requests in all assigned loads.

This means that the AGVs spent more of their time with retrievals for less dense grids. Supposedly, the big number of available escorts for low density grids enabled fast replenishments.

8.5.3 Retrieval Time

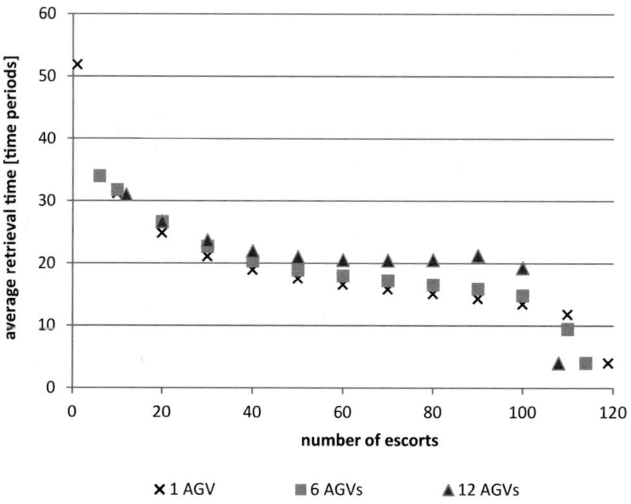

Figure 8.15: Average retrieval time for different numbers of escorts

Figure 8.15 gives the average retrieval time for different numbers of escorts. The average retrieval times decreased with increasing numbers of escorts. Grids with more than 100 escorts showed a differing behavior again. The data suggests logarithmic decline for grids with up to 100 escorts and regression analysis supports this hypothesis (see Table B.3). Therefore, we assume logarithmic decline of the retrieval time for relevant numbers of escorts. The impact of the number of escorts on the average retrieval time seems to be bigger than the impact of the number of AGVs for grids with few escorts. For grids with more escorts, however, the differences between the curves increase, while the slope of the curves is decreasing. Therefore, the impact of the number of AGVs seems to be bigger than the impact of the number of escorts for low density grids (many escorts).

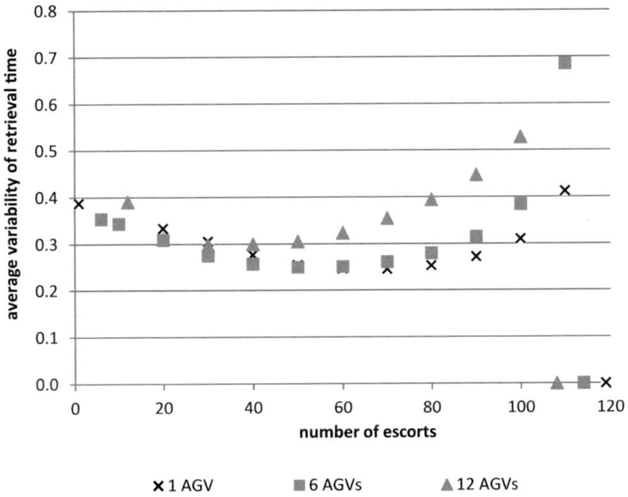

Figure 8.16: Average variability of retrieval times for different numbers of escorts

Figure 8.16 gives the average variability of the retrieval times for different numbers of escorts. Analysis showed a minimum variability for grids with medium density. However, for grids with the individual lowest possible density the variability decreased to 0. Supposedly each AGV was able to move from top to bottom with its request as there were no interfering loads.

8.5.4 Waiting Time and Back Up Moves

Figure 8.17 gives the average absolute number of back up moves, the average number of AGV moves in total and the average retrieval time for different numbers of escorts with 12 AGVs. The other considered numbers of AGVs showed the same structural behavior and, therefore, are not displayed here.

The data indicates that back up moves in average had a minor proportion in the total number of AGV moves. However, there were configurations with high proportions, and the maximum proportion was 80%.

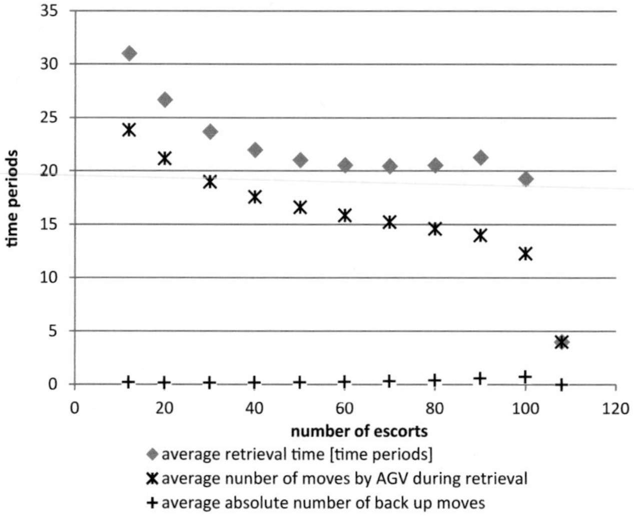

Figure 8.17: Average absolute number of back up moves, average number of AGV moves and average retrieval time for different numbers of escorts with 12 AGVs

Figure 8.17 shows an increasing absolute waiting time for increasing numbers of escorts. The relative waiting time increased with the number of escorts. However, the behavior for the lowest possible density again differed.

Due to the analysis, waiting times may not be neglected.

8.5.5 Summary

We analysed a rectangular grid layout with 30 columns and 4 rows and three different numbers of AGVs. The analysis indicates an increasing throughput with increasing numbers of escorts. The average proportion of requests in all assigned loads supports this hypothesis of increasing throughput as AGVs spend ever more time on requests, which points to a decrease in utilization. Supposedly the bigger number of available escorts reduced the times for retrievals as well as for replenishments.

The data suggests exponential growth of the average throughput with increasing numbers of escorts for multiple AGVs. Thus in high density systems one additional escort will increase the throughput more than proportionally.

In real world applications, there will therefore be a tradeoff of throughput and surface usage which might be controlled by the number of escorts in the grid.

8.6 Influence of the Number of AGVs

To analyse the influence of the number of AGVs, we again used the grid capacity of 120 cells with 30 columns and 4 rows (see Section 8.4). We varied the number of AGVs for this layout with a constant number of escort (12, 24 and 36 escorts).

8.6.1 Throughput

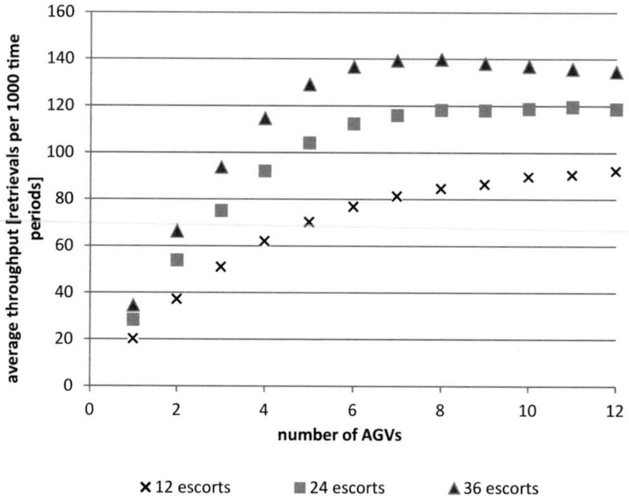

Figure 8.18: Average throughput for different numbers of AGVs

Figure 8.18 gives the average throughput for different numbers of AGVs. For all considered numbers of escorts the throughput increased with the number of AGVs. For 24 resp. 36 escorts the throughput decreased again for more than 11 resp. 8 AGVs. This behaviour is in accordance with the typical behavior of closed networks (Furmans 2000). This system may be regarded to be a closed network as the number of AGVs is constant at all times. The data suggests logarithmic growth and regression analysis supports this hypothesis for 12 escorts (see B.4). For 24 and 36 escorts the coefficients of determination are smaller, supposedly due to the decrease of the average throughput for big numbers of AGVs.

8.6.2 Proportion of Requested Loads in all assigned Loads

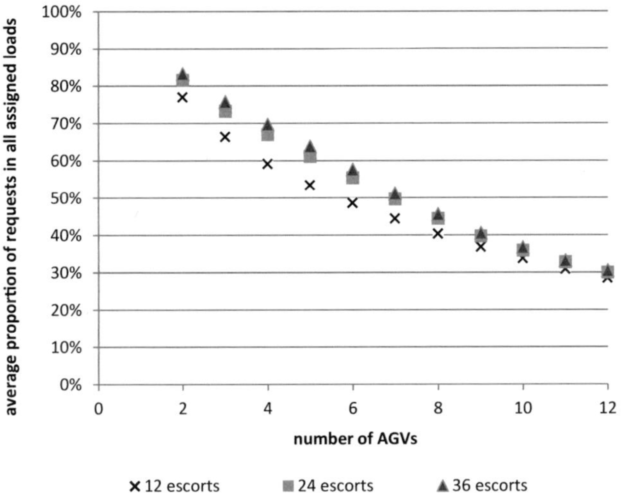

Figure 8.19: Average proportion of requested loads in all assigned loads for different numbers of AGVs

Figure 8.19 gives the average proportion of requests in all assigned loads. The decreasing proportion of requests for increasing numbers of AGVs indicates a high utilization and a possibly imbalanced system for big numbers of AGVs. The results correspond with the stagnating resp. decreasing throughput for big numbers of AGVs.

8.6.3 Retrieval Time

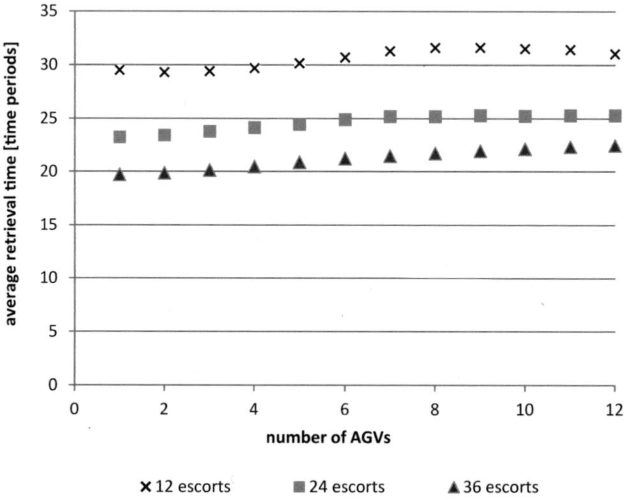

Figure 8.20: Average retrieval time for different numbers of AGVs

Figure 8.20 gives the average retrieval time for different numbers of AGVs. The average retrieval times increased with increasing numbers of AGVs which points to more waiting or blocking for bigger numbers of AGVs.

However, the average retrieval times do not differ more than 10% with the number of AGVs. The average variability of the retrieval times for different numbers of AGVs can be considered to be homogenous.

8.6.4 Waiting Time and Back Up Moves

Figure 8.21 gives the average absolute number of back up moves, the average number of AGV moves in total and the average retrieval time for different numbers of AGVs with 36 escorts. The other considered numbers of escorts showed the same structural behavior and therefore are not displayed here.

The data indicates increasing absolute numbers of back up moves. However, back up moves in average had a minor proportion in the total number of AGV moves. Still, there were configurations with high proportions, and the maximum proportion was 86%.

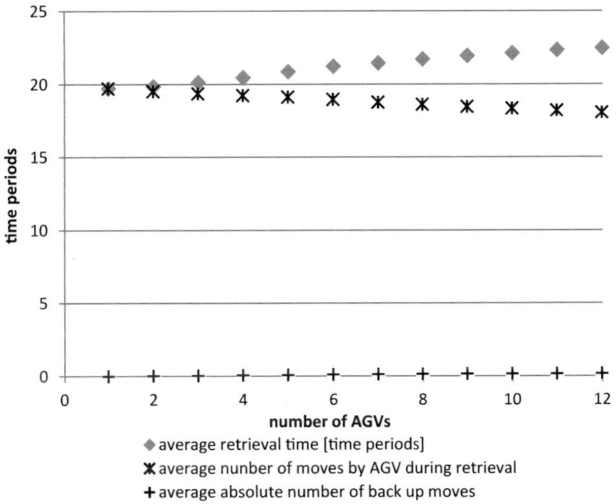

Figure 8.21: Average absolute number of back up moves, average number of AGV moves and average retrieval time for different numbers of AGVs with 36 escorts

Figure 8.21 shows an increasing absolute waiting time for increasing numbers of AGVs as well as an increasing relative waiting time. This behavior is intuitive and in accordance to the stagnating resp. decreasing throughput for a big number of AGVs. If more AGVs are active on the same surface, the probability of two AGVs trying to move into the same cell increases. Thus, waiting times will increase.

8.6.5 Summary

We analysed a rectangular grid layout with 30 columns and 4 rows and three different numbers of escorts. The analysis indicates an increasing throughput with increasing numbers of AGVs. In accordance to the typical behavior of closed networks, the additional throughput per AGV decreases, and, eventually, the additional waiting time outweighs the positive effect of an additional AGV. Therefore, the throughput will stagnate resp. decrease for big numbers of AGVs.

In real world applications there will therefore be a maximum number of AGVs that will still increase the throughput. To use more AGVs will not be useful. For smaller numbers of AGVs there will be a tradeoff of throughput and investment which might be controlled by the number of AGVs in the grid.

8.7 Combined Influence of the Number of AGVs and Escorts

After analysing the influence of the number of AGVs and escorts separately, we also analysed the combined influence of the number of AGVs and escorts. This might give insights on whether adding AGVs and escorts simultaneously is beneficial. For the analysis we again used the grid capacity of 120 cells with 30 columns and 4 rows (see Section 8.4). We varied the number of AGVs and escorts with a constant number of escorts per AGVs. We considered one and two escorts per AGV.

8.7.1 Throughput

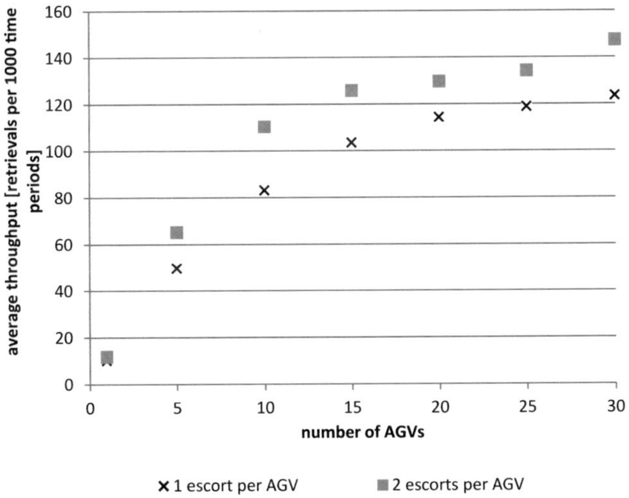

Figure 8.22: Average throughput for different numbers of AGVs and escorts

Figure 8.22 gives the average throughput for different numbers of AGVs and escorts. The throughput increased for increasing numbers of AGVs and escorts. The data suggests logarithmic growth and regression analysis supports this hypothesis (see B.5). Although we analysed higher numbers of AGVs than in Section 8.6 the throughput showed less stagnation. There was no decrease in throughput which indicates that by adding AGVs and escorts simultaneously the throughput is not influenced as strongly as by adding AGVs or escorts separately.

8.7.2 Proportion of Requested Loads in all assigned Loads

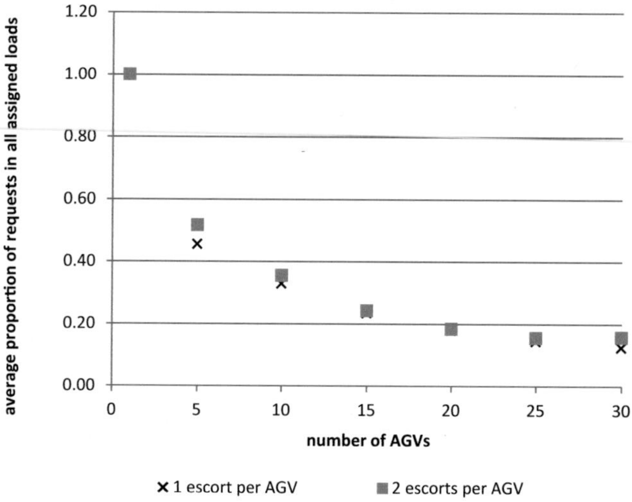

Figure 8.23: Average proportion of requested loads in all assigned loads for different numbers of AGVs and escorts

Figure 8.23 gives the average proportion of requests in all assigned loads. The decreasing proportion of requests for increasing numbers of AGVs indicates a high utilization and a possibly imbalanced system for big numbers of AGVs. Although we considered more AGVs than in Section 8.6, we saw similar minimum proportions.

The results corresponds with the behavior of the average throughput for big numbers of AGVs and escorts.

8.7.3 Retrieval Time

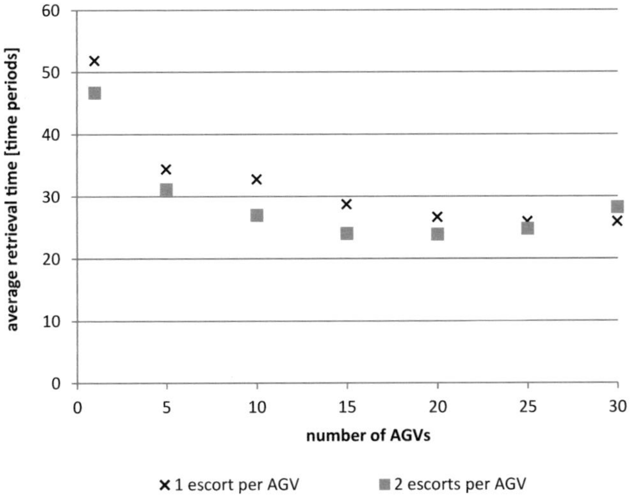

Figure 8.24: Average retrieval time for different numbers of AGVs and escorts

Figure 8.24 gives the average retrieval time for different numbers of AGVs and escorts. The average retrieval times showed a minimum for medium to big numbers of AGVs and escorts. The retrieval time increase for very high numbers of AGVs results from increased waiting (see Section 8.7.4) as more AGVs mean a higher risk of interferences.

The average variability of the retrieval times showed a minimum value for medium numbers of AGVs and escorts.

8.7.4 Waiting Time and Back Up Moves

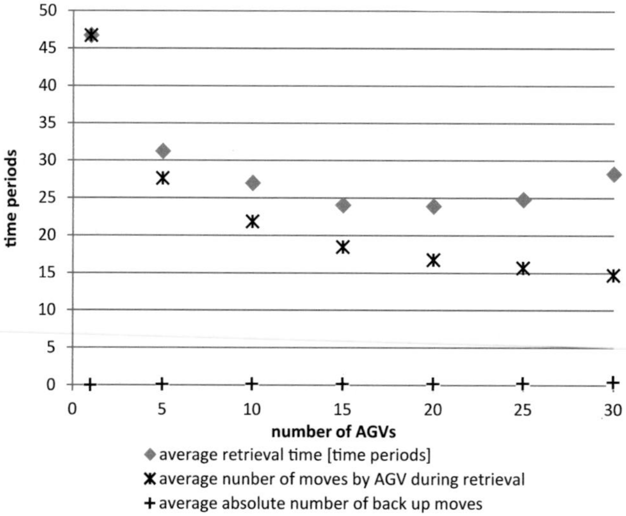

Figure 8.25: Average absolute number of back up moves, average number of AGV moves and average retrieval time for different numbers of AGVs and escorts with 2 escorts per AGV

Figure 8.25 gives the average absolute number of back up moves, the average number of AGV moves in total and the average retrieval time for different numbers of AGVs and escorts with two escorts per AGV. The configuration with one escort per AGV showed the same structural behavior and is therefore not displayed here.

The data indicates increasing absolute numbers of back up moves. However, back up moves in average had a minor proportion in the total number of AGV moves. Still, there were configurations with high proportions, and the maximum proportion was 89%.

Figure 8.25 shows an increasing absolute waiting time for increasing numbers of AGVs and escorts, as well as an increasing relative waiting time. This behavior is intuitive as more AGVs mean a higher risk of interferences and the behavior is in accordance with the logarithmic growth of the average throughput. Although we used bigger numbers of AGVs than in Section 8.6, we saw similar proportions of waiting.

8.7.5 Summary

We analysed a rectangular grid layout with 30 columns and 4 rows, different numbers of AGVs and escort and two different numbers of escorts per AGV. The analysis indicates a similar behavior as in Section 8.6. Still the stagnation of the average throughout is lessened by not only adding AGVs but also escorts.

The analysis suggests to increase the number of escorts along with the number of AGVs in real world implementations. To add escorts with each AGV will however be a tradeoff of throughput and system density.

8.8 Influence of the Number and Positions of I/O Points

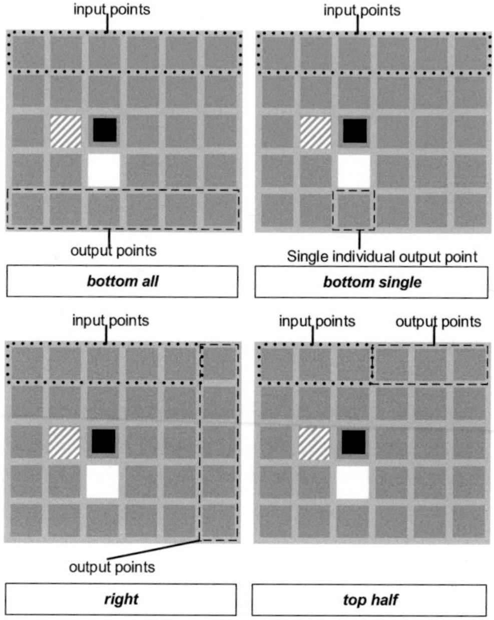

Figure 8.26: Configurations of I/O points

To analyse the influence of the number and positions of the I/O points, we analysed four different configurations of the I/O points (see Fig. 8.26):

- bottom all: input points in all cells in top row, output points in all cells in bottom row
- bottom single: input points in all cells in top row, output point in a single cell in bottom row (individual for each requested load)
- right: input points in all cells in top row apart from rightmost cell, output points in all cells in right column

- top half: input points in left half of all cells in top row, output points in right half of all cells in top row

We used a grid with 12 columns and 12 rows as the configurations of the I/O points involve the grid sides. Therefore, a quadratic layout enables a representative comparison. 9 resp. 16 AGVs and 1 resp. 2 escorts per AGV were used for the analysis.

8.8.1 Throughput

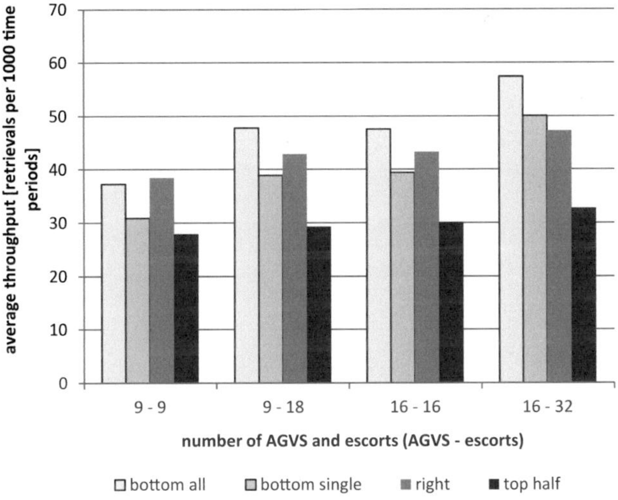

Figure 8.27: Average throughput for different I/O points

Figure 8.27 gives the average throughput for different I/O points. The highest throughput was obtained for the configuration 'bottom all'. This is intuitive as this configuration gives the maximum number of I/O points for each requested load, and there is one main movement direction (down).

The lowest throughput was obtained for the configuration 'top half'. This is intuitive as the number of I/O points is halved in comparison to the configuration 'bottom all' and there are conflicting main movement directions by requested loads and replenishments (up and down).

The order of the configurations 'bottom single' and 'right' depends on the number of AGVs and escorts. In the configuration 'bottom single' a requested load has to reach a single individual output point. In the configuration 'right' there are conflicting main movement directions (right and down).

8.8.2 Proportion of Requested Loads in all assigned Loads

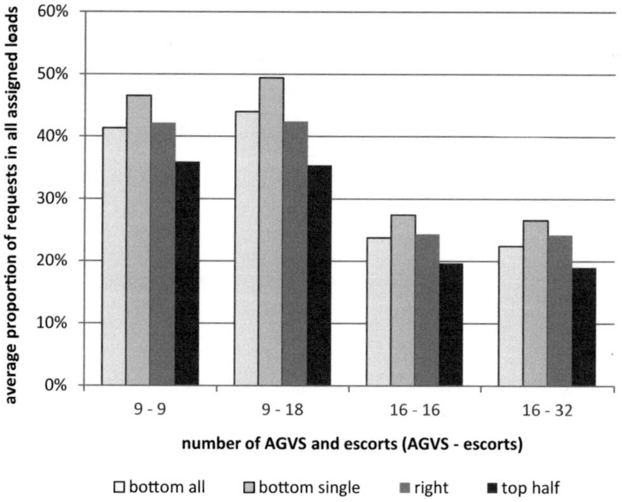

Figure 8.28: Average proportion of requested loads in all assigned loads for different I/O points

Figure 8.28 gives the average proportion of requests in all assigned loads. The configuration 'bottom single' resulted in the highest proportion of requested load. Supposedly the ratio of retrieval time and replenishment time was very high as the requested load has to be moved to a single output point.

The other results are intuitive and correspond with the throughput result.

8.8.3 Retrieval Time

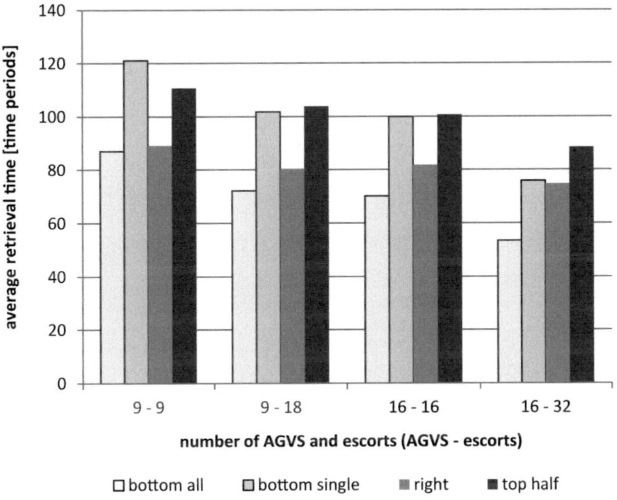

Figure 8.29: Average retrieval time for different I/O points

Figure 8.29 gives the average retrieval time for different I/O points. The results for the configurations 'bottom all' and 'bottom single' correspond to the throughput results.

The configuration 'right' resulted in similar retrieval times to the configuration 'bottom all' for nine AGVs. Supposedly waiting and blocking in the shared grid corner was minor for nine AGVs. For 16 AGVs the configuration 'right' resulted in higher retrieval times as blocking and waiting probably increased.

The configuration 'top half' resulted in big retrieval times. Supposedly blocking and waiting due to the halved number of I/O points and the conflicting main movements direction increased the retrieval times.

8.8.4 Waiting Time and Back Up Moves

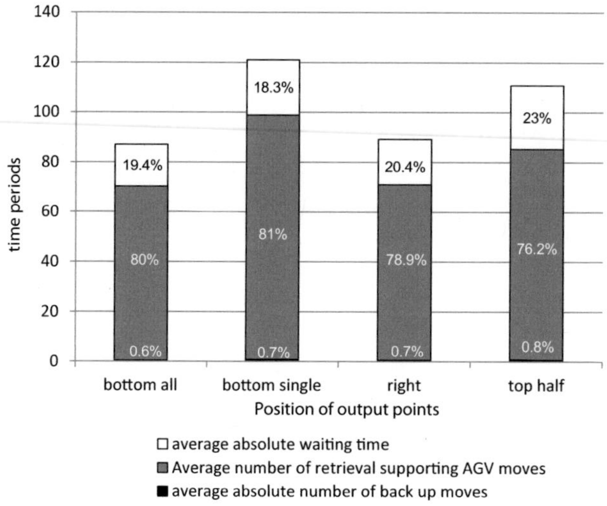

Figure 8.30: Average absolute number of back up moves, average number of retrieval supporting AGV moves and average absolute waiting time for different I/O points with nine AGVs and escorts

other considered number of AGVs and escorts showed the same structural behavior and are therefore not displayed here.

The configurations 'bottom single' and 'top half' showed bigger absolute numbers of back up moves. However, back up moves in average had a minor proportion in the total number of AGV moves for all configurations. Likewise the absolute number of back up moves, the configurations 'bottom single' and 'top half' showed bigger absolute waiting times. These results are intuitive as those configurations contain bigger risks for waiting and blocking as the number of I/O points is smaller and movement directions might be conflicting.

The average relative waiting times are homogenous as the bigger absolute waiting times are divided by bigger retrieval times.

8.8.5 Summary

We analysed a quadratic grid layout with 12 columns and 12 rows, and different numbers of AGVs and escort and four different configurations of I/O points. The configuration 'bottom all' resulted in the highest throughput which is intuitive as this configuration gives the maximum number of I/O points for each requested load, and there is one main movement direction (down). Therefore, the configuration seems to be advisable for real world implementations.

However, there might be reasons to use another configuration.

'Bottom single' might be required for a cross docking scenario if sorting is needed. the configurations 'right' and 'top half' might be used due to layout reasons. However, there will be a tradeoff between layout and surface coverage and system throughput.

8.9 Summary of System Behavior Analysis

To derive insights on the system behavior of GridFlow systems with AGVs and the strategy LivePath, we analysed different configurations with the simulation detail of equal movement times in order not be be influenced by vehicle parameters.

An equal number of AGVs and assigned loads was used in all configurations as vehicles do not cooperate and only handle one request at a time. We wanted to derive insights on the maximum throughput, therefore idle times of vehicles or waiting times of requested loads were not considered. To maintain a constant system density a replenishment was initialized after each retrieval (see Section 4.6.3).

The ensure reliability of the simulation results, the analysis should be based on data from a steady state and the structural influence of the random number generator has to be considered. We successfully verified the results concerning both factors.

The system behavior was analysed for the system parameters grid size, grid layout, number of escorts, number of vehicles as well as number and positions of I/O-points.

Analysis suggests to subdivide big grids to multiple smaller grids in real world implementations as there is a grid size with maximum throughput. However, this grid size depends on the system parameters and may not be obtained generally.

The data points to the advantage of grids with more columns than rows. The optimal ratio of columns and rows however could not be derived in general. Furthermore, in real world applications the chosen layout might depend on other factors such as available space.

The throughput increased with the number of escorts and AGVs. In accordance with the classical behavior of closed networks, the throughput stagnated and decreased for big numbers of AGVs. The data suggests that this effect might be lessened by adding escorts and AGVs simultaneously. For real world implementations there will therefore be a tradeoff of investment in AGVs, surface usage due to system density and the achievable throughput.

The analysis suggests to position the input points and output points at opposite sides of the grid and to use the complete grid edge as input and output points. The throughput decreases if there is a single individual output point or the position and number of input and output points is changed. However, this might be necessary due to the system task and the layout requirements.

The analysis of all system parameters suggests that back up moves in average have a minor importance, but there were configurations with high proportions. In contrast, waiting times may not be neglected as there were considerable average waiting times.

To summarize, the system behavior is well explainable and we were able to derive insights on the behavior as well as on good system design. Due to the not negligible waiting times, there surely is further potential for strategy improvement.

9 GridFlow Case and Influence of Vehicle Parameters

In Chapter 8 we derived insights in the general system behavior of GridFlow systems with AGVs using the assumption of equal movement times (see Section 4.6.3). This means that it takes one time step for an AGV to move to a neighboring cell.

In this chapter we will give insights on the analysis of two specific cases. As we were analysing cases with specific vehicle parameters, we considered the movement characteristics which lead to differing movement times (e.g. loaded or unloaded move, see Section 4.6.3). This enables more realistic results.

Furthermore, we analysed the influence of the vehicle parameters. The following vehicle parameters were considered:

- Loading time: time to pick up the load
- Unloading time: time to put down the load
- Steering time: required time to change the direction of vehicle movement
- Max. velocity (loaded): maximum speed of the vehicle when transporting a load
- Max. velocity (unloaded): maximum speed of the vehicle when moving without a load
- Acceleration (loaded): acceleration factor of the vehicle when transporting a load (deceleration is assumed to have the same value)
- Acceleration (unloaded): acceleration factor of the vehicle when moving without a load (deceleration is assumed to have the same value)

9.1 Modelling and Simulation Details

The movement time for each vehicle to move from one cell to a neighboring cell was individually calculated under consideration of the movement characteristics:

- Is loading and/or unloading required?
- Is there a change of movement direction compared to the last move?
- Is the move loaded or unloaded?
- Is acceleration and/or deceleration required (e.g. if a vehicle moves for several cells without stopping, no acceleration/deceleration will be needed between two cells)?
- Is the full speed reached or does deceleration start beforehand?

As we used a time discrete simulation with the smallest time interval of one second, all movement times were rounded up to full seconds.

To maintain a constant system density, replenishments were considered and the corresponding simulation mode (see Section 4.6.3) was used. Therefore, the maximum throughput of the cases were calculated.

9.2 Buffer Case

We chose a buffer case to analyse the throughput provided by a GridFlow system with AGVs and the strategy LivePath. Additionally we analysed the influence of the vehicle parameters.

9.2.1 Description of Case

First, we analysed a buffer case in a production environment. We assumed parts to be painted in an automated paint job before being assembled. As the lot formation criteria of the paint job and the assembly usually differ (e.g. color in paint job (from light to dark) and part size or diameter in assembly) a buffer might be required. If the number of part variants is small and there are several loads per variant, a supermarket with lanes should be used. However, if there are many variants and few loads per

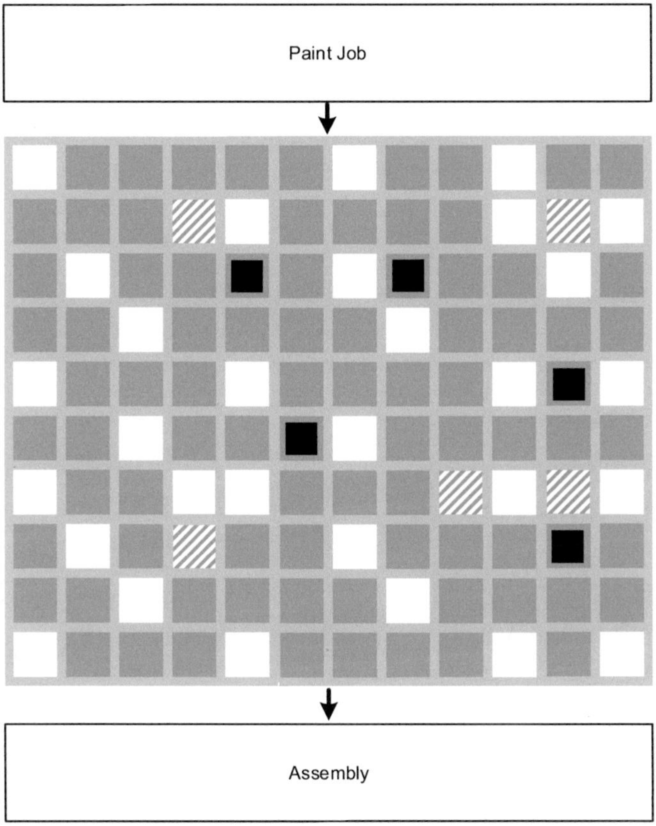

Figure 9.1: System design of the buffer case

variant, a big number of lanes with small lane depths would be required. Additionally, the required lane depths might be differing for the different variants, leading to a high surface utilization as a supermarket with different lane depths is difficult to implement.

In cases like this, a GridFlow systems with AGVs can be implemented to buffer all loads in one grid and retrieve the requested variant when needed.

We used a grid capacity of 120 cells (twelve columns and ten rows), five AGVs and 30 escorts to model the production buffer (see Fig. 9.1). Therefore, a system density of 75% was assumed. All cells in the lower row are considered to be output points for all requested loads, and all cells in the upper row are used for replenishments (configuration 'bottom all', see Section 8.8). The loads are assumed to have the dimensions of euro pallets and to be equipped with a rollable skid. Thus, the vehicles are assumed to move underneath the loads, to connect to the load and to transport it by pulling. Table 9.1 gives the assumed parameters of the vehicles.

Parameter	Value	Unit
Length of a cell (horizontal)	1.1	m
Depth of a cell (vertical)	1.5	m
Loading time	1	s
Unloading time	1	s
Steering time	2	s
Max. velocity (loaded)	0.4	$\frac{m}{s}$
Max. velocity (unloaded)	1	$\frac{m}{s}$
Acceleration (loaded)	0.3	$\frac{m}{s^2}$
Acceleration (unloaded)	0.8	$\frac{m}{s^2}$

Table 9.1: Parameters of buffer case (Nobbe 2015)

9.2.2 Case Results

Table 9.2 shows the results from the buffer case simulation. The system provided an average throughput of 37.8 loads per hour. This means that on average every 1.5 minutes a load was retrieved. This can be considered to be a sufficient throughput of large loads in many buffer scenarios. However, the average relative waiting time of 83% indicates inefficiencies in the system as AGVs spend most of their time waiting. This may be caused by another AGV blocking the cell, the considered vehicle wanted to move to or by an AGV with a higher rank placing a wait request. In the second case, the initiating AGV might have been anywhere in the grid. Therefore, a wait request might also hold an AGV that is not in the neighborhood of the initiating AGV. This supposedly led to inefficient waiting times.

Parameter	Value	Unit
Average throughput	37.8	$\frac{loads}{h}$
Average interretrieval time	1.58	min
Average proportion of requests in all assigned loads	58%	
Average retrieval time	3.83	min
Average number of AGV moves	37.2	moves
Average relative number of back up moves	1%	
Average absolute waiting time	3.2	min
Average relative waiting time	83%	

Table 9.2: Simulation results of buffer case

During active movements, the AGVs spent most of their time with acceleration and deceleration for the considered system and vehicle parameters (see Fig. 9.2). To gain further insights in the influence of the vehicle parameters, we also excecuted a sensitivity analysis.

9.2.3 Sensitivity Analysis of Vehicle Parameters

To measure the influence of the vehicle parameters on the system behavior, we worsened each parameter individually while not changing the other ones. In the analysis, the loading and steering times were doubled and the speeds and accelerations were halved. The unloading time was not considered separately as its influence can be expected to correspond with the influence of the loading time because in average we expect as many loadings as unloadings.

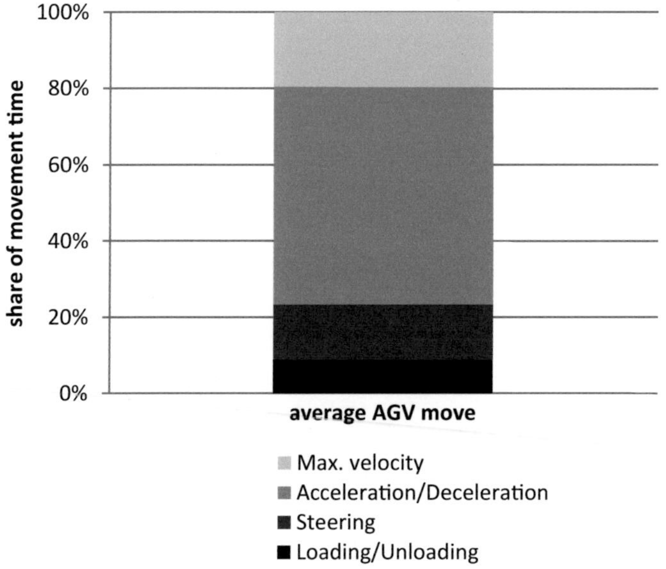

Figure 9.2: Time share of movement time in buffer case

Table 9.3 gives the results of the sensitivity analysis of the buffer case. For all parameters, the worse parameters led to a decreased throughput. The steering time had a bigger impact on the throughput than the loading time. This is intuitive as the time share of steering is bigger.

The speed and acceleration of the loaded moves had a bigger impact than those of the unloaded moves. This supposedly is due to the loaded moves being the bottleneck in the system. Unloaded AGVs will have to wait for loaded ones to finish their longer moves. A decreased speed or acceleration of the loaded move further increases the differences in move

Parameter	Variation	Retrievals per hour	average retrieval time [min]	Average relative waiting time [%]
Loading time	+100%	-6%	+5%	+1%
Steering time	+100%	-11%	+11%	+2%
Max. velocity (loaded)	-50%	-13%	+19%	+4%
Max. velocity (unloaded)	-50%	-5%	+3%	+1%
Acceleration (loaded)	-50%	-17%	+16%	+3%
Acceleration (unloaded)	-50%	-5%	+3%	+1%

Table 9.3: Sensitivity analysis results of the buffer case

times of loaded and unloaded moves. Analysis suggests that this will lead to a bigger impact on the throughput.

This is in accordance with the impact of the parameters on the retrieval and waiting time. The impact on retrieval time and throughput differed from each other as the throughput does not only depend on the retrieval time, but also on the parallelism of retrievals. E.g. with the decreased maximum velocity (loaded) the retrieval time increased by 19%, but the throughput decreased by 13%. Supposedly this is because the proportion of requested loads in all assigned loads increased by 2%. This means that the AGVs spent more of their time with retrievals and less time with replenishments. Supposedly the homogenization of the movement times led to the increase of the proportion of requested loads.

The conclusion will be drawn in Section 10 after analysing the second case.

9.3 Cross Dock Case

Additional to the buffer case in Section 9.2 we analysed a cross dock case.

9.3.1 Description of Case

We analysed a cross docking scenario with five inbound and five outbound docks (see Fig. 9.3). The inbound trucks will unload into the grid and each load will be moved to its individual target dock within the grid, but there will be no sequencing of loads for an outbound truck. The grid has 20 columns and eight rows. Four input resp. output points are assigned to each dock (see Fig. 9.3).

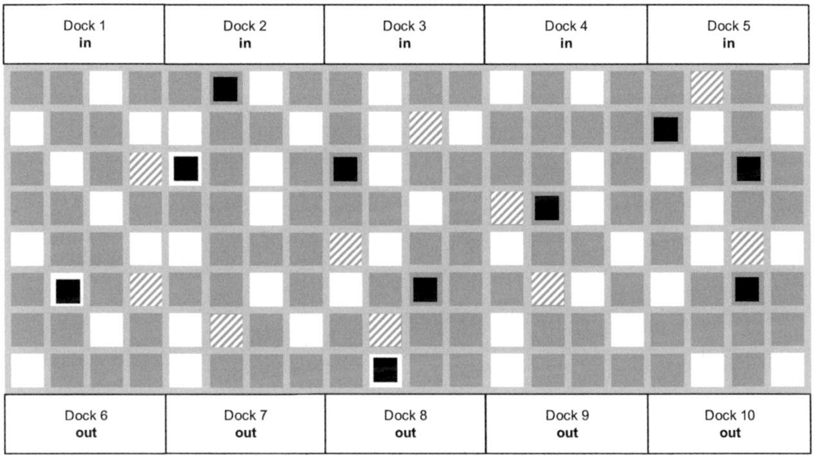

Figure 9.3: System design of the cross dock case

Therefore an inbound truck can use four waiting positions outside the grid to put the loads to. Each requested load has to be transported to one of the output points of its target dock. The next requested load is chosen randomly, and it is randomly assigned to an outgoing dock. If one requested load leaves the grid, a randomly chosen inbound truck will place an ingoing load into one of its waiting positions.

We used a grid capacity of 160 cells (20 columns and eight rows), ten AGVs and 50 escorts to model the cross dock case. Therefore, a system density of 70% was assumed. The loads and vehicles are assumed to have the same characteristics and parameter as in the buffer case (see Section 9.2).

9.3.2 Case Results

Table 9.4 gives the results from the cross dock case simulation. The system provided an average throughput of 35.9 loads per hour. This means that on average every 1.68 minutes a load was retrieved at an outbound dock. Thus, for one individual outbound dock the interretrieval time was 8.4 minutes on average. This may not be considered to be a sufficient throughput of large loads in a short-time cross docking hub as it will take four hours to load an outbound truck with 30 large loads. The throughput might be sufficient for large loads with a longer buffering time within the cross dock.

Similar to the buffer case (see Section 9.2), the AGVs spend most of their time waiting (87%) and back up moves have a minor importance (2.4%).

Parameter	Value	Unit
Average throughput	35.9	$\frac{loads}{h}$
Average interretrieval time	1.68	min
Average proportion of requests in all assigned loads	54%	
Average retrieval time	8.3	min
Average number of AGV moves	59.1	moves
Average relative number of back up moves	2.4%	
Average absolute waiting time	7.3	min
Average relative waiting time	87%	

Table 9.4: Simulation results of the cross dock case

9.3.3 Sensitivity Analysis of Vehicle Parameters

To measure the influence of the vehicle parameters on the system behavior, we did a sensitivity analysis analogous to the buffer case (see Section 9.2). Table 9.5 gives the results of the sensitivity analysis of the cross dock case. The results indicate the same implications like in the buffer case. The steering time, velocity (loaded) and acceleration (loaded) had the biggest impact on the throughput.

9.3.4 Influence of the number of AGVs

As the obtained throughput may not be considered to be sufficient, we also analysed whether additional 10 AGVs would provide a sufficient throughput. With 20 AGVs the system was able to retrieve 41.7 loads per hour. However, this is only equivalent to an increase of 16% and not sufficient for most cross docking scenarios either. Additionally, the average relative waiting time increased to 88%. Thus, the system seems to be close to saturation with 10 AGVs and additional AGVs provide few benefits. Therefore, the strategy LivePath is able to handle cross docking tasks, but as it was not designed for this special purpose, it cannot provide an efficient cross docking process. Major improvements towards to a specialized strategy would be necessary.

Parameter	Variation	Retrievals per hour	average retrieval time [min]	average relative waiting time [%]
Loading time	+100%	-4%	+3%	+1%
Steering time	+100%	-10%	+14%	+2%
Max. velocity (loaded)	-50%	-14%	+17%	+2%
Max. velocity (unloaded)	-50%	-4%	+5%	+1%
Acceleration (loaded)	-50%	-14%	+18%	+2%
Acceleration (unloaded)	-50%	-5%	+5%	+1%

Table 9.5: Sensitivity results of the cross dock case

10 Summary

In today's supply chains, loads spend most of their time waiting in buffers or storage systems. As space in production and logistics facilities becomes ever more limited, efficient buffer and storage systems are needed. Due to investment and flexibility reasons, racking systems might not be the best solution for short-term buffer systems. Block storages on the other hand, hold the disadvantage of limited access to individual loads.

In literature, there are several systems based on the idea of Puzzle-Based Storage Systems, providing a part to picker access to each load. In this system the loads are stored on conveyors or on the ground. The system might be implemented with a grid of locations each equipped with a conveyor or with automated guided vehicles (AGVs) moving underneath the loads in the grid.

First, we developed a taxonomy to classify those systems, proposing the name 'GridFlow systems' to summarize the different systems from literature (see Chapter 3). The taxonomy includes the system task, type of load, technical implementation, system control as well as the movement strategy and system design. Furthermore, we presented advantages of GridFlow systems with AGVs for large loads.

Secondly, we provided the decentralized control strategy 'LivePath' as there is no universal strategy which is able to handle every grid size, number of AGVs, and number of empty locations in literature (see Chapter 4). The strategy is decentralized, i.e. each AGV takes its own decision on the next move based on the current system state. There is no preplanning and reservation of paths, but the path is determined 'live', and only the next cell to move to is reserved and blocked by the AGV. As all AGVs determine their next move individually, there will be situations of conflicts between AGVs blocking each other (deadlocks) or moving in never ending circles (livelocks). Therefore, rules to resolve those situations were implemented and we showed the rules to be able to secure system functionality at all times (see Chapter 5).

Thirdly, to assess the solution quality of LivePath, we presented a method to calculate the optimal movements (see Chapter 6). We used it to compare the optimal solution with the solution gained by the LivePath strategy (see Chapter 6). The strategy LivePath showed promising results, but also further potential concerning the solution quality for a single AGV. However, as the assessment of multiple AGVs was not possible due to solution times, the significance of the assessment is limited.

To derive insights on the system behavior of GridFlow systems with AGVs and the strategy LivePath, we implemented the strategy in simulation software and analysed different configurations regarding the system parameters grid size, grid layout, number of escorts, number of vehicles as well as number and positions of I/O-Points (see Chapter 8). The system behavior is well explainable, and simulation analysis suggests the following implications for good system design:

- Subdivide big grids to multiple smaller grids as there is a grid size with maximum throughput.
- Use grids with more columns than rows.
- Choose the number of escorts and AGVs considering the tradeoff of investment in AGVs, surface usage due to system density and the achievable throughput.
- The benefit of an additional escort or AGV decreased with increasing numbers. Additional AGVs might even decrease This effect might be lessened by adding escorts and AGVs simultaneously.
- Position the input and output points at opposite sides of the grid and use the complete grid edge. The throughput decreases if there is only a single individual output point or the positions and number of input and output points are changed.

Furthermore, we analysed a buffer and a cross dock case. They showed GridFlow systems with AGVs and the strategy LivePath to provide a suitable throughput for production environments, but not for the special task of sorting in cross docks. A specialised strategy derived from LivePath is needed to enable efficient cross docking.

Additionally, there still seems to be a potential for improvement in the buffer case as the AGVs spent most of their time waiting. The results imply the livelock handling to handle more situations than necessary and therefore to cause more waiting than necessary. More efficient loop detection rules might improve the system efficiency.

The existing potential is also supported by the comparison with the optimal solution for single AGVs. To also compare the results for multiple AGVs, a more efficient method to calculate the optimal solution is required.

Sensitivity analysis results regarding the vehicle parameters pointed to short movements with a big share of acceleration and deceleration as well as to frequent steering of the vehicles. To increase throughput, the times to perform these activities could be reduced by technical improvements. Additionally, the throughput could be increased if the strategy could enable the vehicles to travel longer distances with less steering.

To summarize, the strategy LivePath provides the opportunity to implement a GridFlow system with AGVs and decentralized control. General insights on system behavior were be derived and a suitable throughput for a buffer case was obtained in simulation. Further improvement could be achieved in specialising the strategy for the cross docking case and in alterations to decrease waiting times of AGVs.

Glossary of Notation

$A(i,j)$	Adjacency Matrix
$c^2(a)$	variability of indicator a
$c^2_{br}(\overline{tp})$	variability of average throughput $\overline{tp}$ between simulation runs
$c^2_{ir}(tp)$	in-run variability of throughput $tp(t)$
$\hat{c}^2_{ir}(tp)$	maximum in-run variability of throughput $tp(t)$
$C_r(t)$	Cumulated number of retrievals until time period t
$d_{OP}(l)$	Distance of load l to the closest output point
$d_{OP}(l, op_l)$	Distance of load l to output point op_l
$d_{x,OP}(l)$	Horizontal distance of load l to the closest output point
$d_{x,OP}(l, op_l)$	Horizontal distance of load l to output point op_l
$d_{y,OP}(l)$	Vertical distance of load l to the closest output point
$d_{y,OP}(l, op_l)$	Vertical distance of load l to output point op_l
$e \in (1,..,E)$	escort
E	Number of escorts in the grid
$E(a)$	Mean value of indicator a
e_l	escort assigned to load l
$i \in (1,..,N)$	Cell in the grid
$j \in (1,..,N)$	Cell in the grid
$l \in (1,..,L)$	Load in the grid
L	Number of loads in the grid
$m_{v,i,j,t} \in 0,1$	Move by vehicle v from i to j in t
N	Number of cells in the grid
$n_{etc,det}(e, v, nt_l)$	Number of additional moves for vehicle v transport the escort e to the target cell
$n_{etc,dir}(e, v, nt_l)$	Number of moves for vehicle v to transport the escort e to the target cell t_l directly
$n_{etc,total}(e, v, nt_l)$	Number of moves for vehicle v to get to and transport the escort e to the target cell nt_l

$n_{ve}(e,v)$	Number of moves for vehicle v to move to the escort e
$O(l,i)$	Matrix of initial occupancies of cells by loads
$O(v,i)$	Matrix of initial occupancies of cells by vehicles
$o^l_{i,l,t} \in 0,1$	Occupancy of cell i by load l at the end of period t
$o^v_{i,v,t} \in 0,1$	Occupancy of cell i by AGV v at the end of period t
$op_l \in (1,..,OP_l)$	Output point of load l
op^*_l	Closest output point of load l
$p_e \in (1,..,N)$	Position of escort e
$p_l \in (1,..,N)$	Position of load l
$p_v \in (1,..,N)$	Position of AGV v
$R(l) \in 0,1$	Matrix of requested loads
$s_e(a)$	empirical standard deviation of indicator a
$t \in (1,..,T)$	time period
T	Number of time periods
t_b	Neighboring target cell for an AGV having to back up
$t_e \in (1,..,N)$	Target cell for escort e
$t_l \in (1,..,N)$	Target cell for load l
$t_{l,s} \in (1,..,N)$	Target cell for load l for the shortest path
$t_{l,f} \in (1,..,N)$	Target cell for load l for the chosen fast partial path
$tp(t)$	Current throughput in time period t
$\overline{tp}$	Average throughput
$v \in (1,..,V)$	AGV
V	Number of AGVs in the grid
$x \in (1,..,X)$	Column of the grid
X	Number of Columns in the grid
$x_e \in (1,..,X)$	Column of escort e
$x_l \in (1,..,X)$	Column of load l
$x_{op_l} \in (1,..,X)$	Column of output point op_l
$x_v \in (1,..,X)$	Column of AGV v
$y \in (1,..,Y)$	Row of the grid
Y	Number of Rows in the grid
$y_e \in (1,..,Y)$	Row of escort e
$y_{op_l} \in (1,..,Y)$	Row of output point op_l
$y_l \in (1,..,Y)$	Row of load l
$y_v \in (1,..,Y)$	Row of AGV v

References

Alfieri, A., M. Cantamessa, A. Monchiero and F. Montagna (2012). Heuristics for puzzle-based storage systems driven by a limited set of automated guided vehicles. *Journal of Intelligent Manufacturing 23*(5), p. 1695–1705.

Arnold, D. and K. Furmans (2009). *Materialfluss in Logistiksystemen* (6 ed.). VDI-Buch. Berlin and Heidelberg: Springer.

Bartholdi, J. and S. Hackmann (2011). *Warehouse & Distribution Science* (Release 0.95 ed.). Atlanta and GA: The Supply Chain and Logistics Institute, School of Industrial and Systems Engineering, Georgia Institute of Technology.

Furmans, K. (2000). *Bedientheoretische Methoden als Hilfsmittel der Materialflußplanung.* Ph.D. thesis, Universität Karlsruhe, Karlsruhe.

Furmans, K., K. Gue and Z. Seibold (2012). Optimization of Failure Behavior of a Decentralized High-Density 2D Storage System. In: *Dynamics in Logistics, Third International Conference LDIC 2012, Proceedings.*

Furmans, K., N. Nobbe and M. Schwab (2011). Future of Material Handling - modular, flexible and efficient. In: A. Kleiner (eds.), *MMART-LoG at IEEE/RSJ International Conference on Intelligent Robots and Systems (IROS) 2011.*

Gue, K. (2006). Very high density storage systems. *IIE Transactions 38*(1), p. 93–104.

Gue, K. and K. Furmans (2011). Decentralized Control in a Grid-Based Storage System. In: T. Doolen and E. van Aken (eds.), *Proceedings of the 2011 Industrial Engineering Research Conference.*

Gue, K., K. Furmans, Z. Seibold and O. Uludag (2013). Gridstore: A Puzzle-Based Storage System with Decentralized Control. *IEEE*

Transactions on Automation Science and Engineering ; submitted, yet unpublished.

Gue, K. and B. Kim (2007). Puzzle-Based Storage Systems. *Naval Research Logistics 54*(5), p. 556–567.

Gue, K. and O. Uludag (2013). GridPick; http://kevingue.wordpress.com/research/gridflow/gridpick/.

Gue, K., O. Uludag and K. Furmans (2012). A High-Density System for Carton Sequencing. In: Bundesvereinigung Logistik (eds.), *Wissenschaftssymposium Logistik der BVL.*

Martin, H. (2011). *Transport- und Lagerlogistik: Planung, Struktur, Steuerung und Kosten von Systemen der Intralogistik* (8 ed.). Praxis. Wiesbaden: Vieweg + Teubner.

Mayer, S. H. (2009). *Development of a completely decentralized control system for modular continuous conveyor systems*, Volume 73 of *Wissenschaftliche Berichte des Institutes für Fördertechnik und Logistiksysteme des Karlsruher Instituts für Technologie.* Karlsruhe: Univ.-Verl. Karlsruhe.

Mayer, S. H. and K. Furmans (2010). Deadlock prevention in a completely decentralized controlled materials flow systems. *Logistics Research 2*, p. 147–158.

Nobbe, N. (2015). *Vergleich technischer Implementierungen für GridFlow-Systeme.* Ph.D. thesis, Karlsruher Institut für Technolgie, Karlsruhe.

Rohit, K., G. Taylor and K. Gue (2010). Retrieval Time Performance in Puzzle-Based Storage Systems. In: A. Johnson and J. Miller (eds.), *Proceedings of the 2010 Industrial Engineering Research Conference (IERC).*

Seibold, Z., T. Stoll and K. Furmans (2013). Layout-Optimized Sorting of Goods with Decentralized Controlled Conveying Modules. In: SysCon (eds.), *Proceedings of the Systems Conference (SysCon), 2013 IEEE International.*

Taylor, G. and K. Gue (2009). Design and Performance of Multi-Level Puzzle-Based Storage Systems. In: *Proceedings of the 2009 International Conference on Value Chain Sustainability.*

Wenzel, S., M. Rabe and S. Spieckermann (2006). *Verifikation und Validierung für die Simulation in Produktion und Logistik: Vorgehensmodelle und Techniken* (1 ed.). Berlin: Springer Berlin.

List of Figures

List of Tables

A Decision Diagram of a Single AGV in the Grid

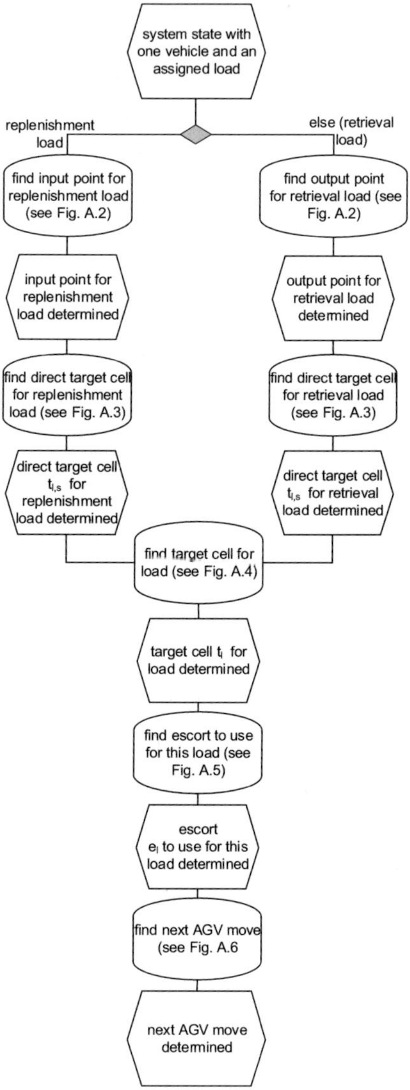

Figure A.1: Decision diagram of a single AGV in the Grid

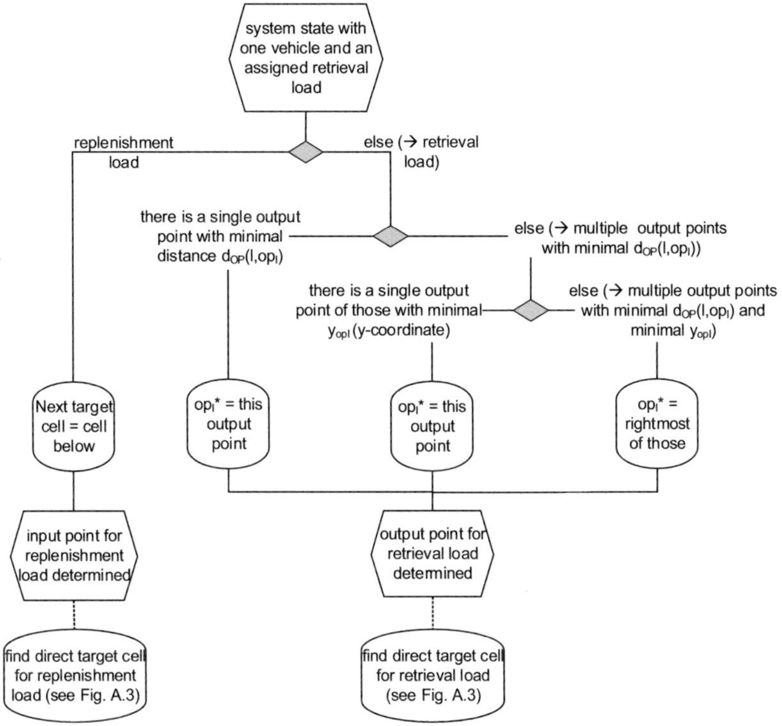

Figure A.2: Decision diagram of a single AGV in the Grid, Step 1a

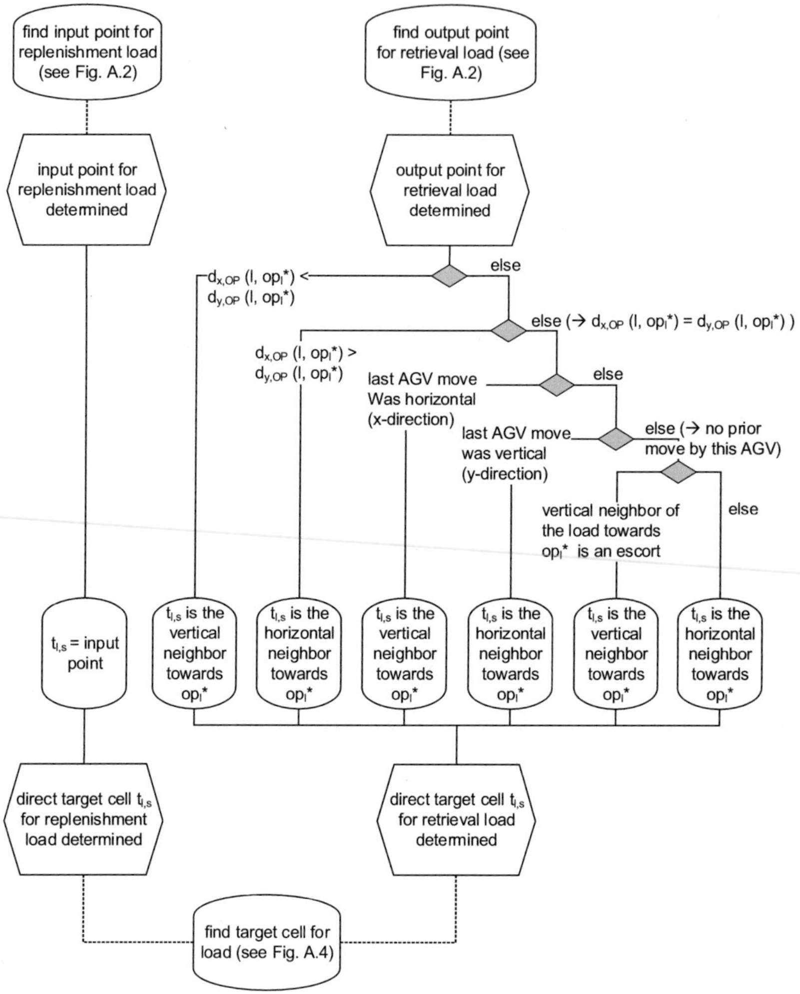

Figure A.3: Decision diagram of a single AGV in the Grid, Step 1b

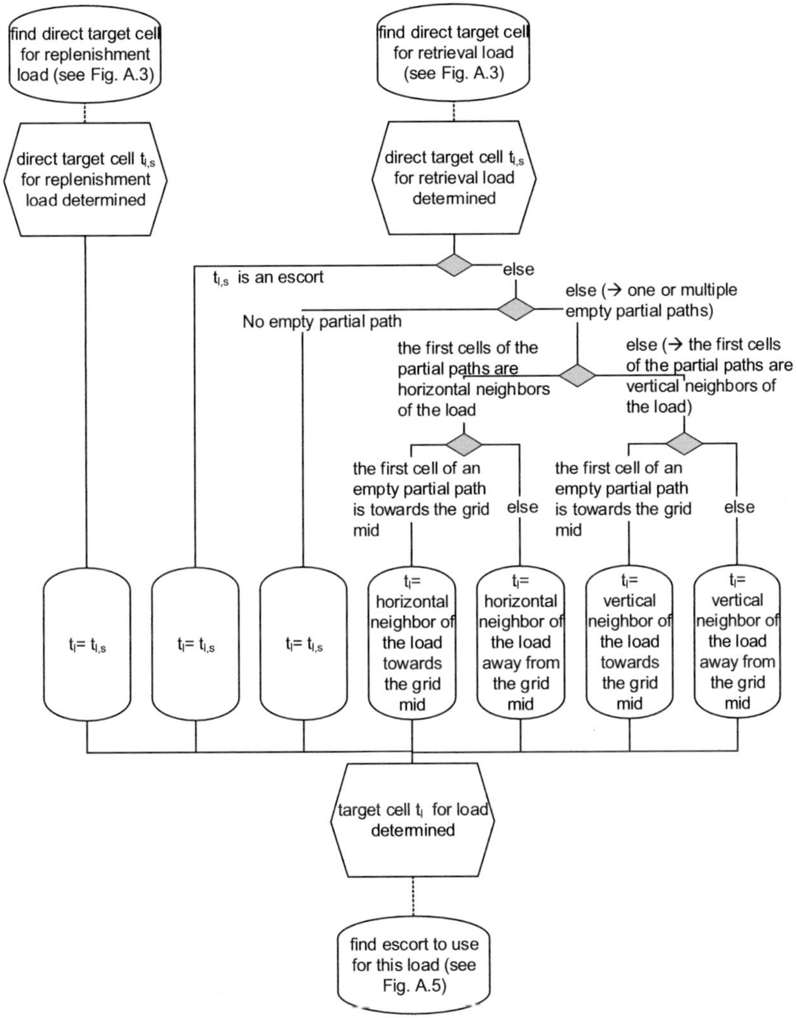

Figure A.4: Decision diagram of a single AGV in the Grid, Step 1c

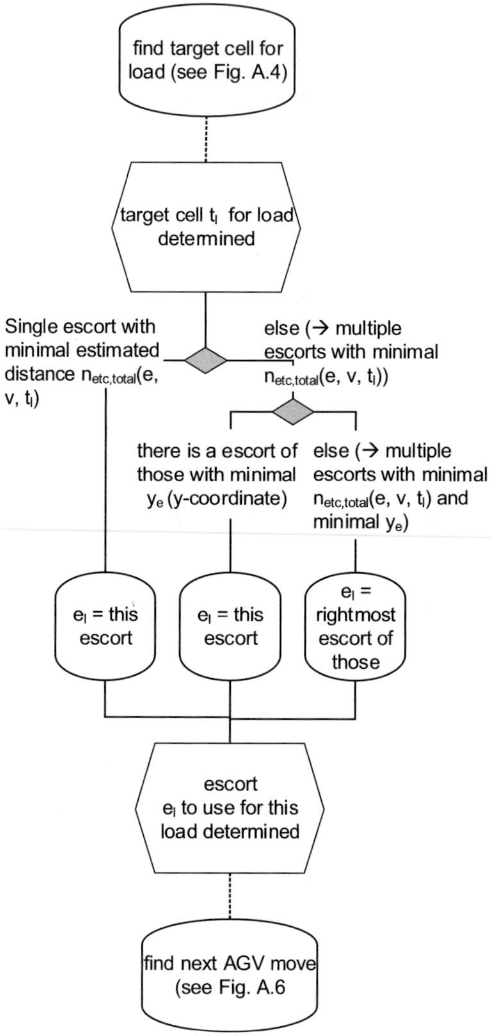

Figure A.5: Decision diagram of a single AGV in the Grid, Step 2

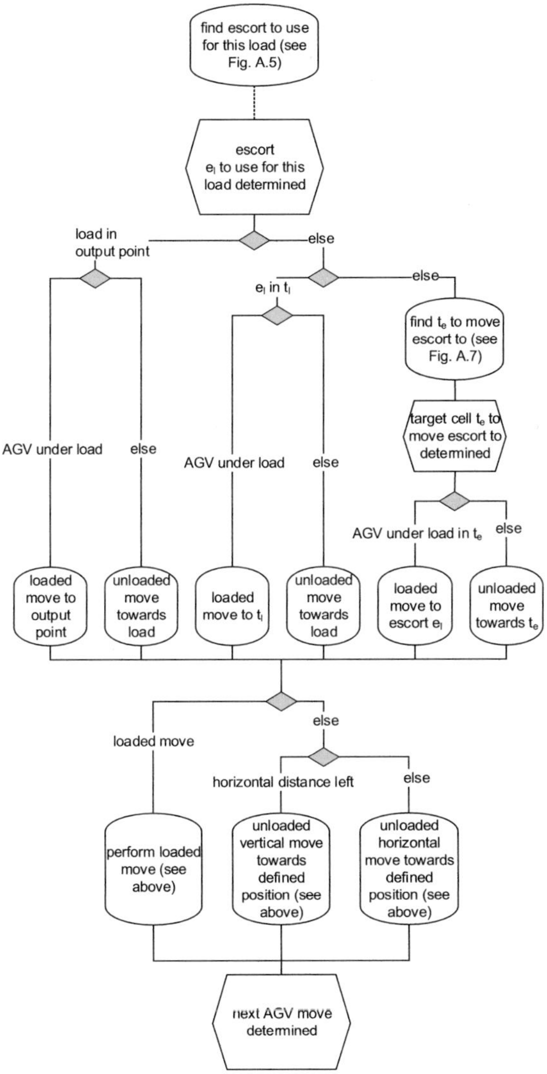

Figure A.6: Decision diagram of a single AGV in the Grid, Step 3a

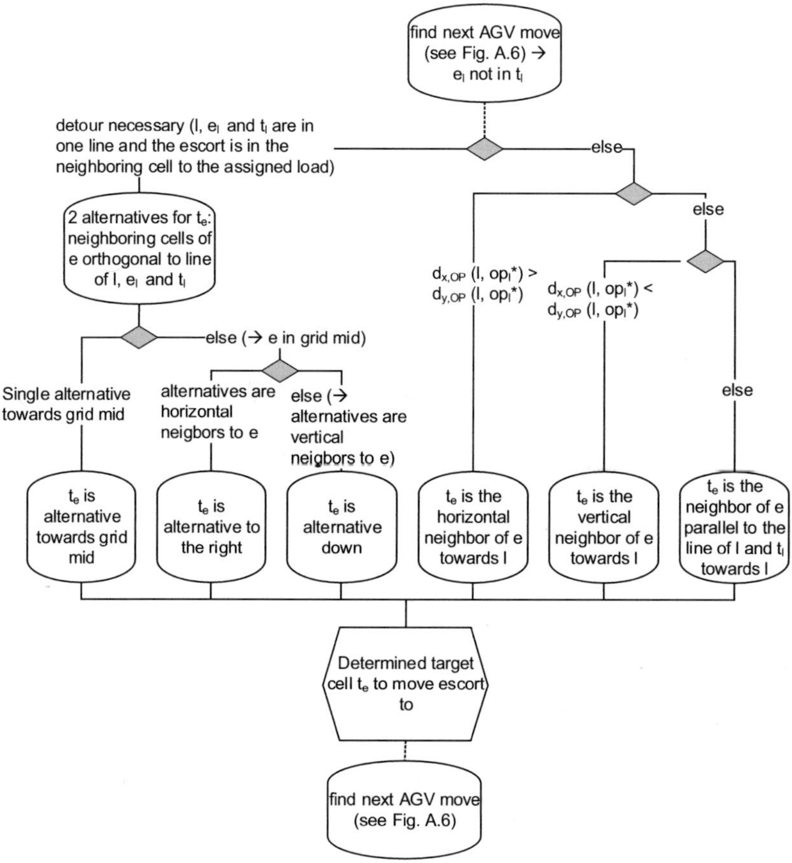

Figure A.7: Decision diagram of a single AGV in the Grid, Step3b

B Data from System Behavior Analysis

	11% AGVs, 11% escorts	11% AGVs, 22% escorts	6% AGVs, 6% escorts	6% AGVs, 12% escorts
coefficient of determination for linear growth of average retrieval time	0.961	0.969	0.961	0.974
coefficient of determination for polynomial growth of order 2 of average retrieval time	0.9974	0.9959	0.9992	0.9989

Table B.1: Regression analysis of grid size

	1 AGV	6 AGVs	12 AGVs
coefficient of determination for hyperbolic decline of average retrieval time (neglecting data for 60 columns)	0.9813	0.9815	0.9708

Table B.2: Regression analysis of grid layout

	1 AGV	6 AGVs	12 AGVs
coefficient of determination for exponential growth of average throughput (up to 100 escorts)	0.836	0.928	0.915
coefficient of determination for linear growth of average throughput (up to 100 escorts)	0.970		
coefficient of determination for logarithmic decline of the average retrieval time (for up to 100 escorts)	0.993	0.997	0.989

Table B.3: Regression analysis of the number of escorts

	12 escorts	24 escorts	36 escorts
coefficient of determination for logarithmic growth of average throughput	0.992	0.950	0.895

Table B.4: Regression analysis of the number of AGVs

	1:1	2:1
coefficient of determination for logarithmic growth of average throughput	0.981	0.984

Table B.5: Regression analysis of the number of AGVs and escorts